GLENN W. SUTER II

The Multi-Media Fate Model:
A Vital Tool for Predicting
the Fate of Chemicals

The Multi-Media Fate Model:
A Vital Tool for Predicting the Fate of Chemicals

Proceedings of a workshop organized by the Society of Environmental Toxicology and Chemistry (SETAC)

Based on an International Task Force which addressed the Application of Multi-Media Fate Models to Regulatory Decision Making

held at

Leuven, Belgium, April 14-16, 1994

and

Denver, Colorado, November 4-5, 1994

Christina E. Cowan

Donald Mackay

Tom C.J. Feijtel

Dik van de Meent

Antonio Di Guardo

Julia Davies

Neil Mackay

Published by SETAC PRESS, 1010 North 12th Avenue, Pensacola, Florida 32501, USA

Direct all inquiries to
Society of Environmental Toxicology and Chemistry
1010 North 12th Avenue
Pensacola, FL 32501-3370
USA

Published by SETAC Press
SETAC Press is an imprint of the Society of Environmental Toxicology and Chemistry.

International Standard Book Number 1-880611-02-3

Printed in the United States of America

♾ The paper used in this publication meets the minimum requirements of the American National Standard for Information Sciences—Permanence of Paper for Printed Library Materials, ANSI Z39.48-1984.

TABLE OF CONTENTS

List of Figures and Tables

Figures

Tables

ACKNOWLEDGMENTS

This workshop would not have been possible without vision and the generous support of these organizations:

- Environment Canada
- Health Canada
- Ministry of Housing, Spatial Planning and Environment (VROM), the Netherlands
- Commission of the European Communities, Directorate General XI; Environment, Nuclear Safety and Civil Protection (EEC, DGXI)
- Association Internationale de la Savonnerie et la Detergence (AIS) and the Comité Européen des Agents de Surface et Intermédiaries Organiques (CESIO)

In addition, many organizations kindly provided support for their representatives to travel to and attend the workshop.

The Organizing Committee was

- Dr. Christina E. Cowan (The Procter & Gamble Company, Ivorydale Technical Center, Cincinnati, Ohio, USA)
- Dr. Donald Mackay (University of Toronto, Canada)
- Dr. Tom C.J. Feijtel (The Procter & Gamble Company, European Technical Center, Strombeek-Bever, Belgium)
- Dr. Dik van de Meent (Laboratory for Ecotoxicology, National Institute of Public Health and Environmental Protection, Bilthoven, the Netherlands)

The Organizing Committee also thanks the following individuals for their invaluable contributions:

- Dr. Antonio Di Guardo (University of Toronto, Canada), Dr. Julia Davies (University of York, UK), Dr. Neil Mackay (Corning Hazleton, UK) for editing the manuscript
- Anne Raë of Procter and Gamble's European Technical Center (Strombeek-Bever, Belgium) for arranging the Leuven workshop
- Lillian Gerde of Procter and Gamble's Ivorydale Technical Center (Cincinnati, Ohio, USA) for arranging the Denver workshop

Finally, the Organizing Committee thanks all the workshop participants who devoted many hours to discussing this topic, writing various sections, and commenting on drafts. Their diligence and hard work were instrumental in ensuring the success of this project.

EXECUTIVE SUMMARY

Mackay-type multi-media models are currently used, and are proposed for use, in regulatory decision making and for assessing the environmental fate of chemicals in several jurisdictions. There has, however, been no systematic evaluation to date of the strengths and limitations of these models. To satisfy this need, the Society of Environmental Toxicology and Chemistry (SETAC) organized two workshops which brought together 28 users and developers of models to discuss applications, to evaluate the performance of different models, to provide guidance on the use of models, and to make recommendations on improvements which will enhance model acceptability and use.

The participants, including representatives who were drawn from government, industry, academic, and consulting groups in Europe, North America, and Japan, met on two occasions: at Leuven, Belgium in April 1994 and in Denver, Colorado, USA in November 1994. At these workshops, the participants split into two groups: users who discussed applications and desirable features of models, and developers who compared models and discussed technical aspects with a view to satisfying the needs of the user group. This report is a summary of the findings from these discussions.

The user group, with the help of the results of a questionnaire which was distributed prior to the workshop, identified several areas of regulatory application and defined the scope and desirable features of models in these areas. The areas identified are

- risk assessment of new and existing chemicals,
- chemical ranking and scoring,
- management of hazardous wastes,
- optimizing testing and monitoring strategies,
- assessment of indirect exposure,
- determining global dispersion and recovery time, and
- testing the coherence of environmental quality objectives (EQOs).

Other application areas including research, education, and life-cycle assessment (LCA) were not discussed.

To provide insights into observed differences in predictions from multi-media models, the model developers conducted a "round robin" comparison of four models: ChemCAN, HAZCHEM, SimpleBOX, and CalTOX. A fifth model, CemoS, was also included, but its difference in structure precluded direct comparison of results. To facilitate comparison, the modelers were provided in advance of the workshop with data on properties and emission rates of 12 chemicals and characteristics of a multi-media environment similar in characteristics to the Netherlands. The independent predictions of chemical fate generated by the models were compared and assessed. Differences between predictions were attributed to two sources. The first arises from different interpretations of specifications and different expressions of units, for example, wet and dry weight-based concentrations. Such differences, although technically trivial, are likely to occur in practice, suggesting that there must be careful, unambiguous definition of all parameters. The second difference arose from the selection of different inter-media transport expressions and parameters such as mass transfer coefficients. During the second workshop, the inter-media transport expressions were adjusted, and it was found that all four models then gave essentially identical results. Comparisons of this type were found to be valuable for establishing model precision and for exchanging information on expressions used to simulate environmental processes. It is recommended that such comparisons be repeated periodically, especially where models are modified or new models introduced.

A number of key issues were discussed and are described in this report, generally resulting in recommendations for model use or further development.

The preferred model is a Level III version, *i.e.*, a non-equilibrium partitioning, steady-state mass balance of chemical fate in at least four media (air, water, soil, and sediments). Simpler Level I and II models may be useful for certain limited applications including gaining an appreciation of chemical partitioning and the importance of chemical reactions. The Level III models are valuable, and even essential, for developing qualitative and quantitative predictions of chemical distribution, pathways, and relative concentrations. They can be used to select and define local site-specific models which will yield quantitative predictions of concentration. The models are well suited for predicting average regional concentrations resulting from highly dispersed and diffused sources such as benzene from car fueling and exhaust. Participants were of the opinion that these models should not be applied to areas smaller than a few hundred square kilometers, nor should they be used for emissions which are highly variable in quantity and periodicity, for example, spills and other intermittent discharges. For certain purposes, unsteady-state Level IV models may also prove useful.

The model should be structured to include air with aerosols, water with aquatic biota and suspended solids, sediments and ideally three soil compartments reflecting differing land types. Detailed models of aquatic and terrestrial food webs, while essential for exposure evaluation, are best treated separately from the principal mass balance model. There was considerable discussion about the desirability of including vegetation as a compartment and improving the treatment of chemical fate in soils and sediments. Suggestions are made for improving models in these respects.

Models should be well documented and transparent. They should not only express the results numerically, but should also quantify the variance in their results which reflects uncertainty. Some form of sensitivity and uncertainty analysis should be included in models to help identify the key input parameters. It is important that the applicability of the model to different classes of chemicals such as nonpolar organics, metals, polymers, and speciating chemicals (phenols, for example) be defined clearly. A tentative chemical classification system is suggested.

Harmonization of different models used in different jurisdictions is attractive from the viewpoint of achieving international consistency of evaluations. One option is to define a single model for use internationally. For several reasons, including the desire not to stifle continuing improvement in models, the need to accept different chemical data sets resulting from different notification requirements, and differences in the use of structure-activity relationships, it was concluded that adoption of a single model is not feasible at this time. Instead, a set of common performance criteria could be defined and models developed to satisfy these criteria. This approach is analogous to defining performance criteria for chemical analyses or instruments.

Models are unlikely to be accepted widely until they are better validated. The issue of validation was discussed at length. Full validation in a scientific sense is probably impossible, but a more modest and attainable objective is to demonstrate that for a variety of chemicals, in defined environments and with adequate emission estimates, the model predictions are in satisfactory agreement with monitoring data. Preliminary results of such exercises are encouraging. There is a compelling case for establishing validation data sets, especially including reliable emission estimates. Recommendations for obtaining such data sets are presented.

The issue of user training and experience was discussed. It is recommended that the greatest effort be devoted to ensuring quality control of the model results and their subsequent use, rather than defining quantifications for the user. Inexperienced users may benefit from using models they do not fully understand.

A number of technical modeling issues were discussed, including the use of Uniform Geographical Units (UGU) as a means of defining regions in terms of groupings of units with common environmental characteristics and the linkage of models to Geographical Information Systems (GIS). The model developers provided a statement of their current perception of the accuracy and precision of multi-media models. They recommended priority research areas which will improve model equations, acceptability, and use.

The wide-ranging discussions of the technicalities and applicabilities of multi-media mass balance models of chemical fate in the environment as recorded in this report provide a comprehensive account of the status of this branch of environmental science and management. It is hoped that it will be useful to the user community, to model developers, and to members of SETAC at large, by providing guidance on how these tools can now and in the future improve stewardship of chemicals which enter our multi-media environment.

1: Introduction

Chemicals may be released into the environment as a result of their normal manufacture, formulation, use, or as a result of their disposal after use. It is important to develop both a qualitative and quantitative understanding of their behavior in the environment. These insights, along with information on possible adverse effects, enable the risk to human health and the environment to be assessed. To achieve this understanding, a chemical's intrinsic properties and emission patterns must be combined with the characteristics of the environment into which it is released to predict how advective and inter-media transport and transformation processes will affect the distribution of the chemical among the various environmental media, and thus the concentrations that will result in each medium. Because the chemical's final distribution and these concentrations are the result of numerous highly complex and interacting processes which are not easily assessed, fate models based on the mass balance principle have been developed to simulate transport among and transformation within multiple environmental media. These models are referred to as *multi-media fate models.*

The most widely recognized and used multi-media mass balance models are those of Mackay, who introduced the concept of fugacity as a convenient method of describing and quantifying the fate of chemicals in air, water, soil, and sediments (Mackay, 1991). The components and key assumptions of these models, which are also called *box models*, are illustrated in Figure 1-1. These models have provided coherence and structure to the field of environmental fate assessment and have made the assessment of the large number of chemicals which are released into the environment more manageable. Multi-media models now exist which have been "translated" from fugacity to concentration units, but all are referred to as *Mackay-type multi-media models.*

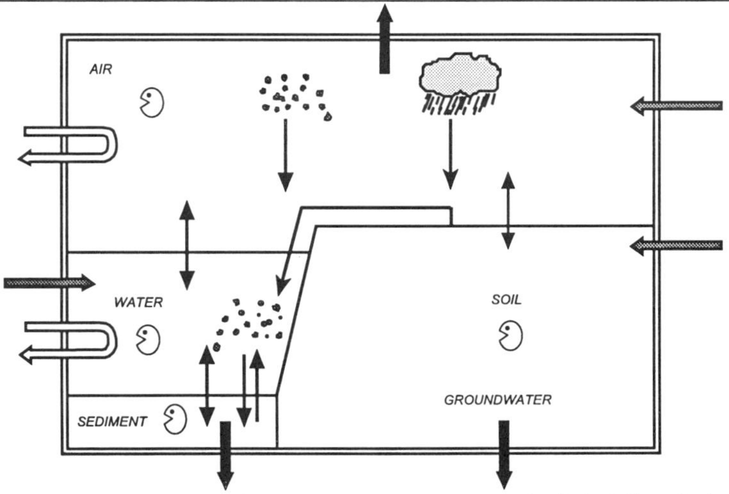

NOTE: Key assumptions include these: (1) media are internally homogeneous and well mixed, (2) partitioning equilibrium is usually assumed within a compartment, (3) rates of inter-media transport are proportional to concentrations. The arrows represent transport processes. Degradation reactions may occur in all media.

FIG. 1-1. Pictorial representation of a multi-media model

Despite the recognition that multi-media fate models are vital and even essential tools for assessing the fate of chemicals released to the environment, there has never been a systematic, international evaluation of their strengths, limitations, precision, and accuracy when used to address the specific needs of regulators and scientists in their various assessment activities. In this report, the term *precision* is used (somewhat unconventionally) to refer to the level of agreement between the results produced by different multi-media models, while *accuracy* refers to the ability of the models to predict actual, measured concentrations in the environment. In fact, when the results produced by multi-media models have been compared (Organization for Economic Cooperation and Development [OECD], 1991), it has been observed that multi-media models from different sources do not always give identical results when applied by different users. The source of these inconsistencies has been attributed, at least in part, to the absence of adequate guidelines for the proper use of the models and/or differences in the expertise of the users. This lack of precision among the various models has raised concerns about their appropriateness in regulatory applications. Furthermore, questions have been raised about the accuracy of the predicted environmental concentrations, and to what extent and under what conditions the environmental concentrations predicted by this type of model can be used in risk assessments. To address these concerns and to provide clarity to the debate on the strengths and limitations of these types of models, the Society of Environmental Toxicology and Chemistry (SETAC) organized two workshops to address the application of multi-media fate models to regulatory decision making and, in particular, to assess and compare the precision of existing models.

The specific objectives of the workshops were to

- identify present and potential future applications of multi-media fate models in regulatory decision making;
- evaluate several multi-media models that are used, or are proposed for use, in regulatory contexts, and identify commonalities and differences;
- provide guidance on precision, accuracy, and applicability of current multi-media models in each area of application; and
- develop recommendations for both future model improvement and applications in regulatory decision-making processes.

This report is a summary of the discussions and deliberations of the 28 participants who were invited to attend the workshops on the "Application of Multi-Media Fate Models in Regulatory Decision Making." These participants included Mackay-type multi-media model developers worldwide, internationally recognized technical experts in the environmental fate of chemicals, as well as expert model users from regulatory agencies, industry, government, and consulting groups in Europe, North America, and Japan. Because the workshop participants are experts in their fields, in many cases the discussion recorded in this report represents expert opinion which in some cases cannot be referenced directly to published reports. The participants hope that this report will be of value by

- providing guidance to users on present and future capabilities,
- identifying steps that can be taken to improve the reliability and usefulness of multi-media models,
- advising model developers on how to ensure model precision, and
- identifying future research needs.

This report is organized into six sections, this introduction being the first. Section 2 describes seven applications of multi-media models. Section 3 describes the results of the model comparison project and discusses harmonization. Section 4 provides guidance on eleven issues relating to model structure and use. In Section 5, the issues of accuracy and validation of the models are discussed. Research needs are identified in Section 6. Following the references, appendices give a list of participants, definitions of some terminology, the questionnaire to users, and suggested criteria for model documentation.

The report is limited to a discussion of Mackay-type multi-media models applied to determining the fate of chemicals in a regional or global environment. It is recognized that there are other types of models that can be called multi-media models (*e.g.*, Multi-Media Environmental Pollutant Assessment System [MEPAS], Droppo *et al.*, 1989) and that multi-media models have been used for other types of chemical fate predictions, including fate in wastewater treatment plants (Struijs *et al.*, 1991), rivers and lakes (Mackay, 1989), and soils (Di Guardo *et al.*, 1994). These models and applications are, however, beyond the scope of this report. Nevertheless, some of the conclusions and recommendations will be applicable to these models. Readers who are unfamiliar with the terminology used to describe these models are referred to Appendix B and to Mackay (1991) for details.

2: Applications of Multi-Media Fate Models

Models ranging from simple dilution equations to complex spatially segmented and temporal variable flow and transport models, including all levels between, are valuable tools for assisting researchers and regulators to understand the fate of chemicals which may be released to the environment as a consequence of routine (*i.e.*, normal manufacturing, use, and disposal) and non-routine (*e.g.*, spills) events. Models are essential when assessing the fate of "new" chemicals that are proposed for use. For existing chemicals, fate models have been used to verify that all processes influencing a chemical's fate and all inputs to the environment have been accounted for, or more generally that our perception of chemical fate is indeed correct. Output from fate models can be used either to obtain a qualitative understanding of the fate of chemicals or to obtain quantitative predictions of environmental concentrations. These environmental concentrations can be compared directly with levels believed to cause effects, therefore giving an indication of risk. They can also be used to predict concentrations in secondary and tertiary compartments for assessing exposure of higher organisms, such as birds and humans, by food ingestion.

To identify applications where Mackay-type multi-media models have been used or are proposed for use, a questionnaire was prepared (Appendix C). This questionnaire was sent to a non-industry target group of model developers and model users. Although the response to the questionnaire by workshop participants gives a good indication of the application needs and future trends, it was not intended to be statistically representative and complete. The results of this exercise, which are summarized in Appendix C, were used to initiate and focus the workshop participants on the main application areas described here.

Mackay-type multi-media models have varying levels of complexity designated as Levels I through III, and occasionally IV, depending on the transformation and transport processes included (Appendix B). The focus of the applications is on Level III multi-media models because there is a consensus among users and model developers that this level of model is required to adequately assess chemical fate. Level I and II models, however, may be useful for developing a systematically more detailed appreciation of the importance of chemical characteristics and identifying which transport and transformation processes determine the chemical's fate (Mackay *et al.*, 1995; Cowan *et al.*, 1995). These simple models will not be applicable generally because significant departures from equilibrium have been observed between compartments. For some of the applications, Level IV or unsteady-state versions of the models may be required to assess the time response of concentrations under dynamic conditions.

Depending on the application, the results obtained from modeling can be used to determine qualitatively how the characteristics of both the chemical and the environment influence a chemical's fate (Mackay *et al.*, 1995a, b, c, d; Cowan *et *al.*, 1995). Generally, these qualitative assessments are conducted using a generic environment (*e.g.*, Mackay, 1991) and generic chemical release scenarios (European Economic Community [EEC], 1993). If the assessments require that the results be used quantitatively, then, as with other types of models, it is critical to have accurate and appropriate information on emissions and on the scale and characteristics of the receiving environments. Table 2-1 shows how the characteristics of the receiving environments corresponding to European countries can vary significantly. The user group identified seven primary areas of application as outlined in Sections 2.2 to 2.8, namely

- risk assessment of new and existing chemicals,
- chemical ranking and scoring,
- management of hazardous wastes,
- optimizing testing and monitoring strategies,
- assessment of indirect exposure,
- determining global dispersion and recovery time, and
- testing the coherence of environmental quality objectives (EQOs).

Non-regulatory application areas, including research and education, were not discussed. The use of these models in life-cycle assessment (LCA) was beyond the scope of this workshop. This issue is under active discussion within the LCA group of SETAC.

2.1 Sequence of Modeling Activities

At this stage, in the interests of clarifying the terminology used and the sequence of modeling activities, it is useful to define the nature of the sequential steps: parameterization, calibration, validation, projection, and post-audit.

2.1.1 Parameterization

Parameterization involves determining the values for the input parameters required for running the model. Parameter values may be defined based on laboratory data or estimation techniques (*e.g.*, quantitative structure-activity relationships [QSARs]). This step helps identify reasonable initial estimates for input parameters and may delineate a range or distribution of values that a given parameter may assume. Lack of information on key model parameters may direct additional research to address such gaps.

2.1.2 Calibration

Calibration refers to the process in which the initial parameter estimates are refined to improve agreement between model predictions and observed data to fall within selected criteria (United States Environmental Protection Agency [USEPA], 1992). This step is not to be viewed as simply an exercise in curve-fitting, but is intended to provide a mechanism for "fine-tuning" the model to site-specific conditions. Obviously, the refined parameter estimates are constrained within the limits defined by the allowable range of the parameter space.

Table 2-1 *Geographical Data Relevant for the Regional Generic Scenario Modeling for European Union Countries*

Country	Number of inhabitants	Area km^2	Water %	Agriculture %	Pasture %	Forest %	Other %	Precipitation mm/y
Belgium+Luxembourg	10,400,000	33,100	0.8	27.1	24.3	21.2	26.6	774
Denmark	5,200,000	43,070	1.6	61.9	7.4	11.0	18.1	800
France	56,900,000	551,500	0.3	33.9	24.9	26.5	14.4	794
Germany	80,400,000	356,910	1.9	34.9	18.1	28.6	16.6	673
Greece	10,100,000	131,990	0.9	29.5	39.8	19.8	10.1	414
Ireland	3,600,000	70,280	2.0	18.3	62.6	3.9	13.3	1030
Italy	58,000,000	301,270	2.4	40.6	17.4	20.7	18.9	773
Netherlands	15,200,000	36,850	8.2	22.6	34.4	8.2	26.5	825
Portugal	9,800,000	92,390	0.5	31.7	5.7	39.4	22.7	920
Spain	39,100,000	504,780	1.0	41.6	22.4	28.9	6.2	520
UK	57,700,000	244,880	1.3	29.3	47.1	8.1	14.3	824
EU total	346,400,000	2,367,020						
EU average*			1.3	35.5	25.7	23.5	14.0	709

* = weighted average
Sources: Number of inhabitants, areas, and precipitations from Hunter, 1992; other data from FAO, 1990

2.1.3 Validation

Validation requires applying a calibrated model to a new scenario and then evaluating the agreement between model predictions and observed data. Validation is an independent test of how well a calibrated model represents the important processes in the natural system (USEPA, 1992). Model validation also provides information regarding the range of model applicability to various problem contexts (*e.g.*, different chemical classes or environmental characteristics). Such information enables the limitations of models to be identified.

2.1.4 Projection

Projection involves the application of a validated model for the purpose of assessing alternate management decisions (*e.g.*, various emission reduction plans).

2.1.5 Post-Audit

Post-audit assesses the accuracy of model projections. Therefore, the post-audit provides a subsequent tier of model validation. For example, a post-audit may involve comparing project reductions in pollutant concentrations to ambient monitoring data after some engineering control is implemented. Although the post-audit is often overlooked, this step provides an important feedback loop for subsequent model validation and refinement.

This discussion provides a general scheme for applying models that are used to predict exposure concentrations. Depending upon the objective of the modeling effort, it is not necessary to apply all steps. For example, multi-media models often may be based only on a parameterization step since field data for calibration and subsequent validation steps are unavailable. Since the reliability of predictions obtained from such calculations is typically unknown, sensitivity or uncertainty analyses should be performed to define the expected distribution of exposure concentrations. This approach requires that the uncertainty of input parameters can be defined or approximated from available data. In addition, this approach assumes that the key processes influencing the fate of the chemical in the environment are included and specified adequately by the model (*i.e.*, the model structure and assumptions are assumed to be correct). These concepts will be discussed further in subsequent sections of this report.

2.2 Risk Assessment of New and Existing Chemicals

Various countries and organizations have enacted legislation requiring assessment of the potential for harmful effects on humans and the environment from the use of new chemicals and from the continued use of existing chemicals. Examples are the Toxic Substances Control Act (TSCA) in the U.S., the Canadian Environmental Protection Act (CEPA) in Canada, the Ministry of International Trade and Industry (MITI) in Japan, and the 7th Amendment in the European Union (EU). Fate models may be used in these assessments to integrate information on the chemical's properties, use and release patterns, quantities, and the characteristics of the environment. They are used to determine the distribution and environmental concentrations of the chemical in the various relevant environmental compartments (*e.g.*, Ghetti *et al.*, 1985; Jori *et al.*, 1986; Devillers *et al.*, 1995).

The nature and quantity of the data provided dictate which models can be used. For the assessment of new chemicals, the amount of physical-chemical and transformation data submitted as part of the chemical's dossier depends on the requirements of the regulatory jurisdiction, and in many cases this is determined by the amount of the chemical that will be used (Table 2-2). In some of these jurisdictions, if the physical-chemical and transformation data required for the fate assessment are not available in the dossier, they may be estimated using quantitative structure-property relationships (QSPRs) or QSARs (*e.g.*, Clements *et al.*, 1993). However, these relationships are available for only a limited set of chemical properties and chemical classes (OECD, 1994). For evaluating the fate of existing chemicals, the physical-chemical and transformation data available are limited to that found in the open literature or provided by the manufacturing or formulating companies. In some cases, there may be a requirement to generate additional data. Thus, the quantity, quality, and type of physical-chemical and transformation data available for assessing the fate of an existing chemical can be highly variable.

Multi-media models have been used or are proposed for use in these types of assessments to provide qualitative and quantitative understanding of the fate and distribution of chemicals. For example, Level III models have been used in assessments produced for the OECD high production volume (HPV) chemicals program to aid qualitative understanding of fate and to estimate environmental concentrations within the Screening Information Data Sets (SIDS) program of OECD. In addition, their use is recommended in the EU technical guidance documents for new and existing chemicals (EEC, 1993; 1994).

More specifically multi-media models do, or can be used to

- identify environmental compartments and the most influential processes that should be evaluated using site-specific models,
- provide estimates of environmental concentrations for assessing the potential for direct and indirect exposure of humans or wildlife,
- identify when advection from neighboring regions may be an important contributor to regional concentrations, and
- provide an estimate of background concentrations for the region of interest.

It must be appreciated that these models estimate average concentrations for the various environmental media in the region. This is a result of the assumption that all media are internally well mixed. Site-specific models which focus on predicting the environmental concentration near the point of discharge are most appropriately used to predict exposure concentrations for site-specific risk assessments, because the highest exposure concentrations will occur in the

TABLE 2-2 *Required Physical, Chemical, and Environmental Testing in Various Regulatory Jurisdictions*

	Europe 7th Amendment		**Canada CEPA**		**Japan MITI**		**U.S. TSCA**
	Base Set	Level I	Schedule II	Schedule III	Notification	Screening†	EPA Expectation
Testing triggers	1 T/year	10 or 100 T/yr	1 T/yr	10 T/yr	1 T/yr	fail Notification test	high volume (100T) consumer use‡
Physical/chemical tests	melting and boiling points density vapor pressure surface tension water solubility K_{OW} flammability etc.	possible additional studies and analytical method developement	melting and boiling points density vapor pressure water and fat solubility K_{OW} and K_{AW} hydrolysis with pH adsorption/desorption		K_{OW} vapor pressure	further phys. chemical tests as appropriate (triggered)	any available data
Environmental toxicology	fish, daphnia, algae acute bacterial inhibition	chronic toxicity terrestrial toxicity BCF in fish	any data available	fish acute daphnia acute		acute fish tox. BCF in fish	3 species acute bacterial inhibition 1 chronic
Environmental fate	ready biodegradable abiotic degradation adsorption/desorption	supplementary fate studies	any data available	ready biodeg.	ready biodeg.		biodegradation studies as appropriate taking into account phys. chem. properties
Comments	Level I test program negotiated based on outcome of Base set testing; Level II negotiated based on Level I results		Some tests may be waived based on scientific justification		no biodegradation triggers "screening" testing on the chemical †slow or incomplete biodegradation triggers "screening" test programs on biodegradation intermediates		‡EPA "chemical categories of concern" may trigger additional tests

vicinity of the release. When releases are limited to only a few point sources within a region, multi-media models have rarely been used in the assessments. The workshop participants recommended, however, that such models be used in these situations as qualitative tools to ensure that all appropriate fate processes are included in the site-specific models. Only when releases are highly dispersed and diffuse (such as benzene emissions from car fueling or exhausts), or when advective processes are important for determining the concentration of chemical in the region, is the multi-media model recommended for quantitatively predicting exposure concentrations. Also of concern for determining the applicability of multi-media models is the time scale and periodicity of the releases. Because the most frequently used version of these models (Level III) assumes constant rates of release and calculates steady-state concentrations, the releases must not be highly variable in quantity or in time.

Because the legislation for new and existing chemicals applies to a wide range of chemicals, the environmental risk assessments and the fate models used in them must be applicable to chemicals with a wide range of characteristics and emission patterns, *i.e.*, to air, water, or soil. Whereas previously, most assessments focused on nonpolar organic chemicals, current needs are for methods to assess the risk of chemicals with unusual properties such as dyes, pigments, and polymers (EEC, 1993). Figure 2-1 illustrates the diversity of uses of new chemicals which had been notified to the EU by January 1993 (EEC, 1993).

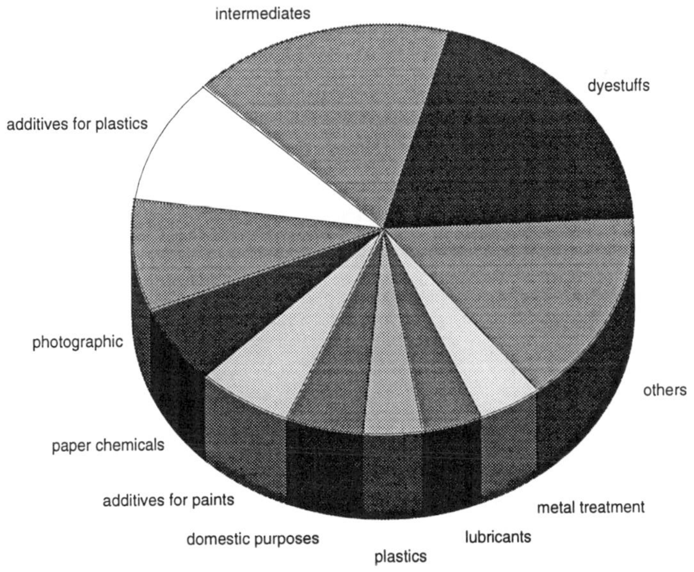

FIG. 2-1. New chemicals notified to the EU by January 1993

Most regulatory systems operate with a tiered approach. As a consequence, different amounts of data will be available in different stages of the notification process. Thus, at lower tiers, only estimated data may be available for input to the models, while at the highest level, a full range of measured data may be available. The participants recommended that if data beyond that available at the tier of interest are required to run a model, then a consistent and transparent method must be provided to obtain default and/or estimated values. They also recommended that these models be capable of using the full range of data available at each tier.

Major causes of uncertainties in the predictions should be identified and quantified. This will include assessment of uncertainties in the input data. The participants thus recommended that the models be capable of estimating the uncertainty in the predicted environmental concentration associated with uncertainty in the input data.

2.3 Chemical Ranking and Scoring

The large number of existing chemicals (between 60,000 and 100,000 in commerce) implies that there is a need to set priorities in order to identify those which should be assessed first. Although scaling and ranking systems are still being developed, and SETAC has an active program of providing guidance on a scientifically based approach, it is envisioned that future systems will include some information on fate and exposure in addition to the current emphasis on effects. Within this context, there may be a need for simple fate models and the ability to account for multi-media partitioning on the chemical's fate and distribution. The models should seek to identify the media in which the highest concentrations or highest potential for impact will occur.

One current prioritization system that includes a fate component is that of the EU's priority setting method, Informal Priority System (IPS) (Bureau, 1994). This system will be used to prioritize chemicals for more in-depth evaluation of risk. The IPS method ranks chemicals on the basis of their potential danger to man and the environment.

In assessing the potential danger to the environment, this method includes a simple exposure-effect model. A Mackay Level I model with six compartments (water, air, soil, sediment, suspended solids, and fish) is used to calculate the environmental distribution of the chemical at equilibrium and steady-state.

Because the environmental ranking is defined in relationship to the chemical's potential to affect the aquatic environment, the fraction calculated to be in the water compartment is used as one of the parameters to generate the Aquatic Priority Score (APS). The environmental characteristics are as described in Annex III of van der Zandt and van Leeuwen (1992) and in Bureau (1994). The uncertainty in the ranking of a chemical is evaluated using the range in the input data. Although this is a very simple way of incorporating fate information into priority-setting, this approach demonstrates one use of multi-media fate models in such schemes.

Ideally, multi-media fate models used in ranking schemes should be able to predict the fate of chemicals with a wide range of characteristics and emission scenarios. The range of existing compounds that are to be ranked include metals, organo-metals, and metalloids as well as hydrophobic organic chemicals. Because physical-chemical and transformation data on many existing chemicals are limited, the model should require as few chemical properties as possible. Furthermore, it is desirable to be able to estimate these properties (using QSPRs or QSARs, for example) when data are missing. Because the fate of a chemical is dependent on its characteristics, including use and emission pattern, the participants felt that ideally the ranking should take these characteristics into account. This ranking should also focus on identifying in which environments the chemical will achieve highest concentration, and not necessarily on predicting the actual exposure concentrations. For this reason, the environmental characteristics should be generic, *i.e.*, identical for all chemicals evaluated. Furthermore, the participants requested that the multi-media fate model that is used in the ranking, although simple, should be descriptive enough to simulate with reasonable accuracy the distribution of chemicals in the environment. It is recommended that information be provided on the uncertainty of a given rank.

2.4 Management of Hazardous Wastes

Another very common application of fate models is to understand the fate of chemicals that are in, or may migrate from, hazardous waste sites. This information can be used to determine if there is any potential for adverse effects on humans or the environment in the vicinity of the waste site. A subsequent step is to determine clean-up criteria and to help identify the most effective remediation method. Because of the small size of waste sites and the limitations of multi-media fate models to represent processes and concentrations on a very small site-specific scale, multi-media fate models are not generally recommended for use in most of these types of assessment. The exception to this is a multi-media fate model such as the CalTOX model (McKone, 1993) which is used at the screening level to assess the exposure to human populations in the vicinity of the hazardous waste site from waste that is buried in or has contaminated the soil. For this very specific type of application, there are several justifications for the use of a multi-media fate model. First, the focus is on screening and thus on predicting the amount of the chemical that may be transported off site. Secondly, the source of the chemical is usually below the surface soil. For this scenario, the multi-media model can be applied to a smaller scale with a recommended smallest size of 1,000 m^2. In general, it is recommended that if multi-media models are to be used to evaluate hazardous waste sites, they be designed for this purpose. Other fate models can be used which address the heterogeneity of the site and the contamination (*e.g.*, MEPAS, Droppo *et al.*, 1989). In specific situations where the multi-media model approach is not sufficiently accurate to support regulatory decision, and where it is necessary to develop a site- and case-specific model, a multi-media model may be used as a first step to identify which processes the site-specific model should address in more spatial detail.

2.5 Optimization of Testing and Monitoring Strategies

Fate models can assist in the design of relevant and cost-effective testing strategies for chemical properties and effects. Experience has shown that for a variety of chemicals, not all of the properties (solubility in water, log K_{OW}, vapor pressures, *etc.*) are of equal importance for predicting the chemical's distribution and concentration in environmental compartments (*e.g.*, Mackay *et al.*, 1992). This implies that when fate models are used to predict these distributions and concentrations, accurate values are needed for some properties and less accurate estimates are acceptable for others. There is no general rule that can be applied to determine when a property should be determined with greater accuracy. For example, if a chemical has a very low vapor pressure (*i.e.*, 10^{-8} atm) and is discharged into a soil with high organic matter content, it is unlikely to evaporate. Log K_{OW} will be an important parameter and vapor pressure a less important parameter for determining this chemical's fate. At low values of log K_{OW}, vapor pressure will become more important.

To determine the most influential properties for a chemical, the model is first parameterized with available data and/or estimated values for the properties (*e.g.*, based on experience or QSPRs). A sensitivity analysis of the output for ranges in the property values will indicate which have the greatest influence on the fate of the chemical. This is typically performed by perturbing each input parameter individually by a small percentage (*e.g.*, 1%) and determining the extent to which the predicted concentrations change in response. Chemical property testing can then focus on accurate experimental determination of the most influential properties, and estimated values may be adequate for others. This procedure can be used for any fate model and is particularly applicable to multi-media models where the distribution among environmental media varies greatly depending on the particular set of parameter values chosen for the chemical (*e.g.*, OECD, 1991; Mackay *et al.*, 1992).

The concept of *value of information* can be applied in this context. Models are often useful as tools to assess the value of new information. The economic criterion is that new information should be collected only when the financial cost of collecting new information is balanced by the value of that information. Value of information in a regulatory context is measured by the extent to which that information will change the regulatory decision. Thus, if there is an inaccurate value of K_{OW}, and we are likely to make a decision to ban a chemical because we assume it has a high K_{OW}, collecting data that will improve the accuracy could change this decision. Thus, this new information has a high value. In contrast, if the decision to control exposure is not sensitive to uncertainty about the true magnitude of K_{OW}, then improving the accuracy has little value.

Fate models in general, and multi-media models in particular, can also be used to focus effects-testing on the most critical environmental media and the types of organisms in those media that are most likely to be exposed. The output of a multi-media model will indicate those environmental compartments with the highest chemical concentrations, allowing the effects-testing to be focused appropriately. For example, if a high concentration of the chemical is found in sediment, then testing of sediment toxicity should be given high priority.

For existing chemicals, the results of both qualitative and quantitative assessments can be used to identify the environmental compartments of most interest and thus guide future monitoring studies. For example, predictions of fate and concentration of a substance in the environment can be used to design an optimal monitoring study by focusing sampling and analytical effort on those compartments predicted to exhibit the highest concentration.

Monitoring can also be used to verify the fate assessment and fate model predictions. Thus, there should be a fundamental interplay between monitoring and modeling.

2.6 Assessment of Indirect Exposure

For human health and wildlife risk assessments, both direct and indirect environmental exposure must be assessed; the pathways by which chemicals migrate through the environment and into media from which exposure may occur are illustrated in Figure 2-2.

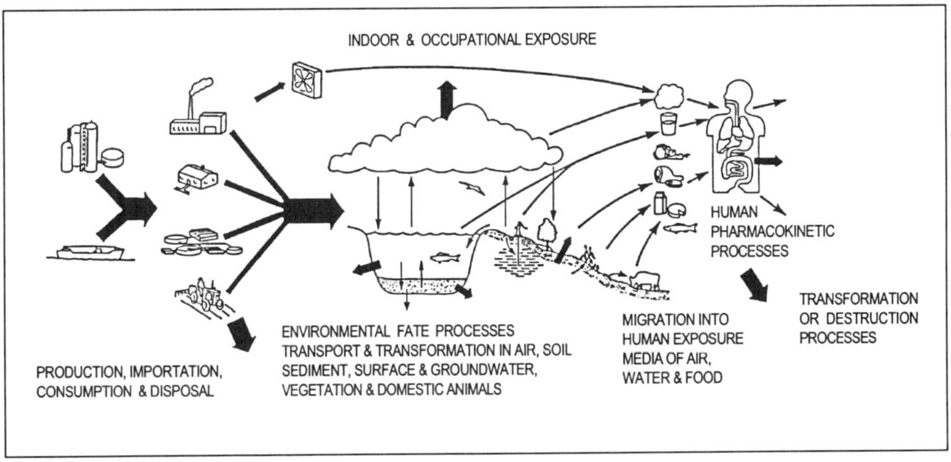

Fig. 2-2. Illustration of movement of chemicals through the environment and into media, resulting in human exposure (modified from Mackay, 1991)

The results from multi-media fate assessments can be used directly if the exposure is due to contact with the primary environmental compartments of air, water, soil, ground water, and sediment; or indirectly if exposure occurs via secondary and tertiary compartments. Secondary compartments include biotic compartments which receive and/or accumulate the chemical from the primary environmental compartments. Generally, the most important secondary compartments are fish and plants. If the chemical is further transported through food chains to humans, *e.g.*, grass to cows to meat/milk to human, then tertiary compartments must also be included in the exposure assessment. The final step in assessing the exposure is to estimate the quantities taken up from the primary, secondary, and tertiary exposure compartments, such as intake of "food basket" components for different target groups (*e.g.*, age groups, feeding habits).

Recently there has been increasing interest in evaluating indirect exposure by food chain transport and effects on top predators and wildlife. For example, the Dutch National Institute of Public Health and Environmental Protection (RIVM) developed a method to calculate the concentration in food of top predators (fish and earthworms as food for birds-of-prey and mammals) from the concentration in water and soil (Jongbloed *et al.*, 1994). The procedure for performing these assessments is similar to that used for assessing indirect exposure in humans.

Multi-media models can play a significant role in these assessments by predicting average concentrations in the primary exposure compartments for the region of interest. An average regional concentration for the primary compartments is generally more appropriate than the higher point-of-entry concentrations because the resulting exposure, especially through secondary and tertiary compartments, represents an average over the region of interest. The multiple secondary and tertiary exposure pathways can be included into or added onto the multi-media models by including the chemical's transfer from primary to secondary compartments. Because the exposure concentrations in the secondary and tertiary compartments are dependent on the concentration in the primary compartment, an accurate quantitative estimate of the concentration in each of the primary exposure compartments is critical to the accuracy of these indirect exposure assessments. A limited number of nonpolar organic chemicals and a few organo-metals have been shown to bioaccumulate to the extent that secondary and tertiary exposure routes are significant (Young *et al.*, 1981; Young, 1988; Suedel *et al.*, 1994). The most valuable application of multi-media models will therefore be to poorly metabolized lipophilic chemicals.

In those cases where the focus of the assessment is on a local site and specific population, it may be necessary to develop a detailed site-specific model encompassing the biological, chemical, and physical processes and exposure patterns of the receptor population. This approach would be warranted when evaluating the exposure to human or wildlife populations in the vicinity of a specific waste site. In these cases, the multi-media model can be used to estimate background concentrations and to indicate which media and processes should be included in the site-specific model. Examples of such site-specific models are presented in Jørgensen (1990, 1994).

2.7 Determination of Global Dispersion and Recovery Time

Nested multi-media models can be used to determine how far and how fast on a global scale a chemical will disperse as a result of local releases. These models account for advection of a chemical from the regional to continental and to global scales. The potential for a chemical released locally to disperse to continental and global environments is related to the relative rates of advective and transformation losses. At RIVM, "nested" multi-media models have been developed as shown in Figure 2-3. Another approach attempted by Wania and Mackay (1995) has been to compile a meridionally segmented global model as shown in Figure 2-4. In this approach the compartments are air, agricultural soil, non-agricultural soil, freshwater, freshwater sediments, and surface ocean water. The air and ocean compartments are linked to adjacent meridional compartments. Nested and meridional multi-media models can also be used to estimate the time needed for eliminating a specified percentage of the chemical from the environment after cessation of emission. This time is referred to as *recovery time* and is related to the chemical's persistence. Persistence on a regional scale is determined by the combined rates of the reaction and advection losses; however, on a global scale, persistence is the result of reactive losses only.

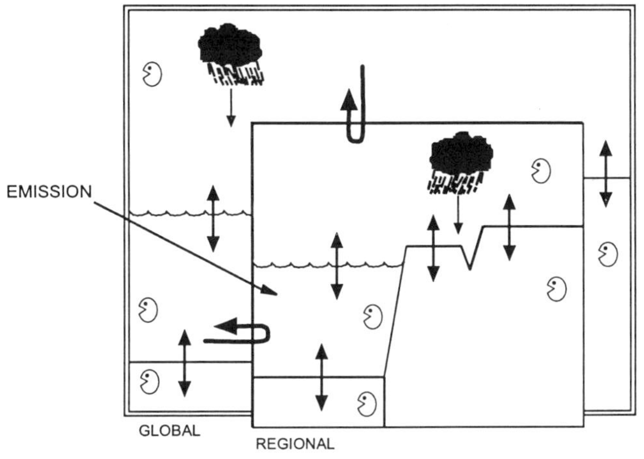

Fig. 2-3. Nested multi-media model

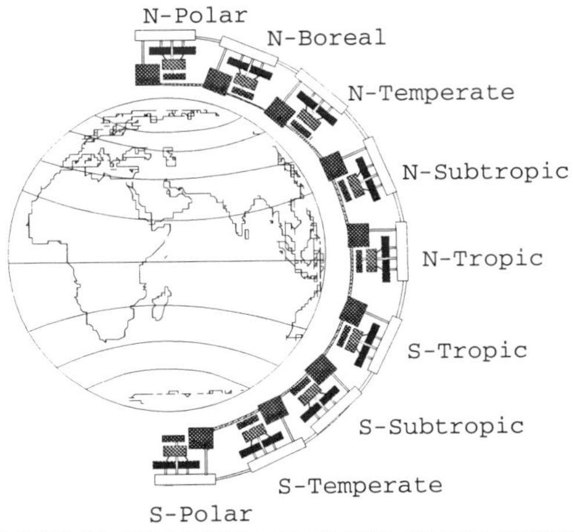

Fig. 2-4. Meridional multi-media model

A first approximation of recovery time on a regional scale can be the *residence time* calculated by a Level III multi-media model. This residence time is the quotient of the total amount of the chemical present at steady-state in the environment and the sum of the rates of input (or output, since at steady-state inputs and outputs are equal) to that environment. More precise estimates of recovery time will require a Level IV multi-media model to predict the concentration in all environmental compartments over prescribed periods of time.

2.8 Testing Coherence of Environmental Quality Objectives

Another unique application of multi-media fate models is that of van de Meent and de Bruijn (1995), who used multi-media models to test for the coherence or congruence of EQOs. An environmental quality objective (*e.g.*, a water quality criterion) is the concentration at which there will be no adverse effects. EQOs have typically been established for each environmental compartment independently (*i.e.*, for air, water, and soil) without taking into account the possible transport of a chemical into other compartments. One method of assessing whether these objectives are consistent is to set the concentration in one of the compartments to its EQO and calculate the resulting steady-state concentration from inter-media transport in neighboring compartments using a multi-media steady-state model. When the concentration in these other compartments exceeds their EQO, then the EQOs are not coherent. The extent to which the EQOs must be adjusted to render the entire set consistent can also be determined using the multi-media models. In principle, such analyses could be conducted for all types of chemicals for which criteria have been proposed in more than one environmental medium.

It is noteworthy that one reason for developing the CalTOX model was to help establish generic objectives for the clean-up of soils. These objectives were to be dictated by the concentrations and resulting exposure in other compartments as determined by the extent of inter-media transport.

2.9 Common Needs

The following several common needs can be identified from these diverse applications of models:

- **Types of chemicals.** The models must be suitable for a wide range of chemical types. They should be able to handle chemicals which can exist in different forms (*e.g.*, metals in different oxidation states, ionizing chemicals, non-aqueous liquids, and non-soil associated solids such as pure naphthalene).
- **Types of release.** The models should be able to predict the distribution and environmental concentrations for several types of emission scenarios, including both continuous and intermittent releases to different compartments.

- **Types of input data required.** For most of the applications, the amount and quantity of physical-chemical and transformation data available for a chemical is unknown and likely to vary among chemicals. The one exception is new chemical evaluations where the required data are prescribed in the regulations.
- **Simplicity versus complexity.** Because the amount of available data will vary, simple models are preferred. They should have the ability to increase in complexity as the data and application warrant.
- **Uncertainty analysis.** Systematic methods are required for calculating the variability in the model output due to the variability in chemical parameters, emission rates, and/or environmental characteristics.
- **Sensitivity analysis.** Systematic methods are required to determine which chemical parameters and environmental characteristics have the greatest influence on a chemical's fate.

3: Multi-Media Model Comparison

To provide insights into differences between multi-media models that have been used or are proposed for use in regulatory applications, the model developers conducted a comparison of four of these models. This exercise compared model structure, including the compartments and multi-media processes that are contained in the models. Prior to the workshop, the model developers compared the predictions from these multi-media models to determine if they would give consistent results when applied to the same simulations. This was essentially an evaluation of model precision. The results and findings from these comparisons were used as the basis for the recommendations and discussions presented in later sections of this report.

3.1 Preliminary Evaluation of the Models

Whether equilibrium, steady-state, or transient, all multi-media models are based upon the law of conservation of mass. In 1789, Antoine Lavoisier showed that a "balance sheet" method was required to account for the quantities of chemicals involved in complex chemical and physical transformations as observed in the laboratory. Multi-media models are natural extensions of this "balance sheet" method, but in this case the time and length scales have been extended to include decades and centuries and larger regions or even the entire Earth. All the multi-media models focus on single chemical species and exploit the now accepted axiomatic statement that mass is conserved in and among the environmental compartments.

The models included in the evaluation were as follows:

- **CalTOX** is a fugacity-based model that combines Level IV soil-layer compartments with a steady-state Level III model containing four compartments. The CalTOX model was developed to assist the California Environmental Protection Agency to estimate chemical fate and human exposure in the vicinity of hazardous waste sites. CalTOX also includes indirect exposure algorithms and tools to quantify and assess uncertainties in model outputs. (McKone, 1993)
- **ChemCAN** is a steady-state Level III model, developed for Health Canada, which uses the fugacity and aquivalence approach to predict a chemical's fate in any of the 24 regions of Canada. The model contains four main environmental compartments and a number of sub-compartments. It is very similar to the equilibrium criterion (EQC) series of evaluative models (Mackay *et al.*, 1995b) which deduce chemical fate in a hypothetical 105 km^2 region and to the basic model described by Mackay *et al.* (1992). A version is available as a Windows application. (Mackay *et al.*, 1995c)
- **HAZCHEM** is a steady-state Level III model with five compartments, two of which are soils. It is based on the same fugacity model as ChemCAN but contains a number of modifications. It uses matrix inversion to find the steady-state solution. It was developed as a regional scale model for all EU

Member States. (European Centre for Ecotoxicology and Toxicology of Chemicals [ECETOC], 1994a)

- **SimpleBOX** is a steady-state Level III model. It consists of eight compartments, three of which are soils of differing use and properties. It also produces quasi-dynamic (Level IV) output by using an external numerical integrator. Because the model was developed at RIVM as a regional scale model for the Netherlands to be used as part of the Uniform System of Evaluation of Substances (USES), its default environmental characteristics represent the Netherlands. SimpleBOX uses the classical concentration concept to compute the mass balances. (van de Meent, 1993)

A fifth model, CemoS, developed at the University of Osnabrück, was also included in the comparison but it proved to be quite different in concept from the others, thus direct comparison of model output is difficult. It can be viewed as a "tool box" of model modules.

3.2 Similarities and Differences in Model Structure

The structures of the models are summarized and compared in Table 3-1. All models use similar well-accepted expressions for describing chemical transformation and transport and can yield estimates of non-equilibrium concentrations in multi-media well-mixed "box" environments under steady- and in some cases unsteady-state conditions. Examination of the model structures reveals that the SimpleBOX, HAZCHEM, and ChemCAN models are very similar Level III multi-media models. HAZCHEM is very similar to ChemCAN because it used the source code from Mackay *et al.* (1992) as a starting point for the model development. HAZCHEM includes minor modifications to the model formulation regarding the representation of sedimentation and resuspension and in the way that residence time in air and water are estimated.

SimpleBOX is conceptually nearly identical to ChemCAN and HAZCHEM, because they are based on the same theory. However, SimpleBOX treats suspended matter and biota as separate compartments that are not necessarily at equilibrium with water. These compartments can be forced to equilibrium by setting the time constant for equilibration to (almost) zero. Furthermore, SimpleBOX does not follow the fugacity (Z- and D-values) approach, but instead uses concentrations, mass transfer coefficients, and inter-media partition coefficients. In addition, some of the transfer mechanisms, especially for sediment deposition, resuspension, and capacity of the aerosols, are represented differently among ChemCAN, HAZCHEM, and SimpleBOX. These differences in representation represent technical areas in which further research is needed.

CalTOX is also a Level III multi-media model with formulations similar to those of the other models. However, it differs significantly from the other models in the way that it treats soil and by the fact that it includes vegetation as a separate compartment in the model. Rather than assuming a single soil layer for different land uses, CalTOX includes a three-layer soil column.

Table 3-1 *Comparison of the Structure of Selected Multi-Media Models*

Model Descriptors	CalTOX	ChemCAN	HAZCHEM	SimpleBOX
Formulation	Fugacity	Fugacity	Fugacity	Concentration
Numerical Solution Method	EXCEL	Explicit analytical solution	Matrix inversion routine	Compiled Lotus 123 spreadsheet
Chemical Domain	Type 1 and 2	Type 1 and 2	Type 1 and 2	Type 1 and 2
Default Environments	Residential, Commercial, and Industrial Landscapes in California	24 Canadian Regions plus Generic	10 European Regions and European Generic	Netherlands
Envir. Compartments				
Air with Aerosols	yes	yes	yes	yes
Soil	1 land use type, w/3 vert. layers	1 soil use type	2 soil use types	3 soil use types
Sediment	yes	yes	yes	yes
Water	yes	yes	yes	yes
Aquatic Biota	no	subcomp. of water	subcomp. of water	sep. compartment
Suspended Solids	subcomp. of water	subcomp. of water	subcomp. of water	sep. compartment
Terrestrial Vegetation	yes	no, included in add-on indirect exposure model	no, included in add-on indirect exposure model	no, included in add-on indirect exposure model
Diffusive Inter-media Transfer Processes				
Soil/Air	volatilization and gas absorption	volatilization and gas absorption	volatilization and gas absorption	volatilization and gas absorption
Water/Air	volatilization and gas absorption	volatilization and gas absorption	volatilization and gas absorption	volatilization and gas absorption
Water/Sediment	sorption and desorption	sorption and desorption	sorption and desorption	sorption and desorption
Water/Aquatic Biota	none	equilibrium partitioning	equilibrium partitioning	based on 1st order kinetics
Advective Inter-media Transfer Processes				
Soil to Air	resuspension of soil particles	none	none	none
Air to Soil	wet and dry deposition	wet and dry deposition	wet and dry deposition	wet and dry deposition
Air to Water	wet and dry deposition	wet and dry deposition	wet and dry deposition	wet and dry deposition
Soil to Water	soil-solution run-off and erosion	soil-solution run-off and erosion	soil-solution run-off and erosion	soil-solution run-off and erosion
Water to Sedim.	sediment deposition	sediment deposition	sediment deposition	sediment deposition
Sedim. to Water	sediment resuspension	sediment resuspension	sediment resuspension	sediment resuspension
Other	uptake by vegetation from soil and air	uptake by vegetation from soil and air	none	none

TABLE 3-1 — *Continued*

Model Descriptors	CalTOX	ChemCAN	HAZCHEM	SimpleBOX
Transform. Processes in Air, Water, Soil, Sed.	overall 1st order rate	overall 1st order rate	overall 1st order rate	overall 1st order rate
Loss Processes from Regional Environm.				
From Air	advection	advection	advection	advection
From Soil	leaching to groundwater	leaching to water	leaching to water	leaching to groundwater
From Water	advection	advection	advection	advection
From Sediment	burial	burial	burial	burial

3.3 Round Robin Model Comparison

The second part of the model comparison involved comparing the output of the four models in a "round robin" exercise to explore the ability of the models and modelers to give consistent model outputs, given the same or similar input (*i.e.*, to evaluate model precision). The scenarios that were distributed to each of the model developers included

- a table of physical-chemical properties and media-specific half-lives for nine chemicals which spanned a wide range of properties (Table 3-2);
- specific emission rates, *i.e.*, 1000 kg/hr into specific media only (air, water, and soil) and 1000 kg/h simultaneously into air, water, and soil; and
- a multi-media environment or landscape of defined dimensions and properties (Table 3-3).

The model developers were asked to calculate the steady-state amounts and concentrations in the compartments, residence times, and certain other response times. Inter-media transport parameters such as mass transfer coefficients were not specified, but were left to the modeler to select. The results of the questionnaire (Appendix C) and the model comparisons were compiled and presented to the participants to initiate discussion.

It is appropriate here to note that the selection and critical evaluation of input data as in Table 3-2 is a demanding task. Often, reported parameter values cover a range of values, and temperature must be taken into account with proper averaging. Particularly difficult is the selection of reaction half-lives, especially when only laboratory data are available. For example, extrapolation of a laboratory biodegradation rate to environmental conditions in soils, water, or sediment is problematic. Since the results are very dependent on the selected values, considerable effort should be devoted to the selection process. The reader is referred to handbooks of compiled data such as those by Howard (1989) or Mackay *et al.* (1992a, b, 1994, 1995) for further discussion of this issue.

Initial comparison of the results from the various models revealed some unexpectedly large discrepancies which proved to be from two sources. Some of

Table 3-2 *Chemical Properties for the "Round Robin" Assessment*

Chemical	MW g/mol	MP °C	SOL g/m3	VP Pa	K_p L/kg	Log K_{ow}	HL in air h	HL in water h	HL in ground-water	HL in soil	HL in sediment
Trichloroethylene	131.39	-73	1100	9900		2.53	100	550	550	1700	5500
Benzo(a)pyrene	252.3	175	0.0038	7E-7		6.04	170	1700	1700	17000	55000
1,4-Dichlorobenzene	147.1	53.1	83	90.2		3.4	550	1700	1700	5500	17000
Hexachlorobenzene	284.2	230	0.005	0.0023		5.5	30000	55000	55000	55000	55000
Hexachlorobiphenyl (PCB 153)	360.9	103	0.01	1.2E-4		6.9	30000	50000	50000	100000	100000
Pentachlorophenol	266.45	174	14	0.00415		5.05	550	550	2000	5000	5000
Atrazine	215.69	174	33	3.85E-5		2.75	5	20000		2000	1000
Linear Alkylbenzene Sulfonates (LAS)	348		62.5		20 (soil) 600 (sed)			24	10000	500	100
Lead					500000 (soil & sed)		∞	∞	∞	∞	∞

Note: MW = molecular weight; MP = melting point; SOL = solubility in water; VP = vapor pressure; K_p = soil-water partition coefficient; K_{ow} = octanol-water partition coefficient; HL = half-life

Sources: Mackay *et al.*, 1992a, b, 1994

the discrepancies were attributable to differences in interpretation of the specified input and environmental characterization data and to the use of different units such as wet versus dry weight-based concentrations.

TABLE 3-3 *Dimensions and Properties of the Multi-Media Regional Environment (from ChemCAN, used with permission from Health Canada)*

Regional Data

Type of Scenario: User defined   Region: Comparison

Dimensions of Environment		Volume Fractions for Subcomp.	
Total surface area (km²)	50000	Particles in air	2E-11
Surface covered by water (% of total)	10	Particles in water	.00001
Length of coastline for maritime regions (km)	0	Fish	.000001
Average air height (km)	1	Air in soil	.2
Average water depth (m)	3	Water in soil	.3
Average soil depth (cm)	10	Soil solids	.5
Average sediment depth (cm)	3	Sediment pore water	.8
Average coastal water depth (m)	10	Sediment solids	.2

Temperature Conditions (°C)	
◉ Year round mean	20
○ Winter mean	10
○ Summer mean	29

Fraction Organic Carbon	
Particles in water	.05
Soil solids	.05
Sediment solids	.05

Residence Times (d)	
Air	.5
Water	50

Add Region | Save Region | Delete Region | Print | Cancel | OK

The most important differences arose because inter-media transport data were selected independently by each modeler, and this resulted in large discrepancies for the chemicals with extreme physical-chemical properties (for example, trichloroethylene [TCE]). Some of the results of this model comparison exercise are shown in Table 3-4, which gives the amounts present at steady-state in air, water, sediment, and soil, as predicted for the case of simultaneous emission to the three compartments (*i.e.*, soil, water, air). The results generally appear to agree, with only a few exceptions. Similar steady-state distributions were predicted by the different models for most of the chemicals. There were differences in the predicted absolute amounts (proportional to concentrations). The median coefficient of variation was 67%. For air, the agreement was judged to be satisfactory for all chemicals except linear alkylbenzene sulfonates (LAS) and lead, for which ChemCAN predicted lower concentrations then did the other models. For water, the maximum difference among the models was typically of a factor of 10. For sediment, larger differences (up to a factor of 50) were reported. For soil, the agreement was much better for most chemicals with the exception of lead, for which the difference between SimpleBOX and CalTOX amounted to a factor of 110.

Table 3-4 *Steady-State Masses (kg) Predicted for Nine Chemicals (emission 1000 kg/h to air, water, and soil)*

Compartment	Model	TCE	B(a)P	1,4 DCB	HCBenz	HCBiphen	PCP	Atrazine	LAS	Lead
AIR	CalTOX	2.94E+04	4.17E+03	3.23E+04	3.07E+04	1.80E+04	8.61E+03	7.73E+02	1.22E+04	1.23E+04
	ChemCAN	2.65E+04	8.15E+03	2.40E+04	1.76E+04	1.19E+04	1.13E+04	9.54E+02	1.43E+02	1.61E+02
	HAZCHEM	3.16E+04	8.84E+03	3.02E+04	2.28E+04	1.30E+04	1.15E+04	8.94E+02	2.09E+03	8.72E+03
	SimpleBOX	3.19E+04	9.26E+03	2.97E+04	2.37E+04	1.55E+04	8.19E+03	2.64E+03	1.17E+04	1.04E+04
	coeff of var	0.08	0.27	0.13	0.25	0.20	0.17	0.59	1.36	0.83
	max factor diff *	1.2	2.2	1.3	1.7	1.5	1.4	3.4	85.4	76.4
WATER	CalTOX	1.85E+05	2.56E+05	2.22E+05	2.12E+05	1.39E+05	2.21E+05	1.93E+06	3.57E+04	3.09E+05
	ChemCAN	2.67E+04	4.33E+04	2.94E+04	3.04E+04	3.27E+04	4.63E+04	6.17E+04	2.27E+04	9.26E+04
	HAZCHEM	5.45E+04	1.21E+05	6.52E+04	6.29E+04	4.38E+04	2.92E+05	1.38E+06	3.67E+04	9.34E+04
	SimpleBOX	1.64E+04	2.33E+05	1.70E+04	2.79E+04	2.11E+05	4.34E+05	2.01E+06	3.44E+04	1.07E+06
	coeff of var	2.40	0.75	2.54	2.16	0.88	0.63	0.78	0.21	1.11
	max factor diff *	11.3	5.9	13.1	7.6	6.5	6.3	32.6	1.6	11.6
SEDIMENT	CalTOX	1.16E+04	4.61E+07	6.45E+04	9.75E+06	6.72E+07	3.76E+06	8.61E+04	7.78E+03	1.83E+08
	ChemCAN	8.59E+02	8.30E+05	3.73E+03	2.10E+05	1.68E+06	7.96E+04	1.59E+03	5.73E+02	6.78E+06
	HAZCHEM	1.96E+03	9.07E+06	1.45E+04	1.66E+06	9.53E+06	1.72E+06	4.26E+04	2.62E+03	2.62E+07
	SimpleBOX	1.26E+03	4.85E+07	1.27E+04	2.53E+06	4.66E+07	4.34E+05	1.32E+04	7.81E+02	3.13E+08
	coeff of var	3.79	1.27	2.67	2.90	1.61	2.24	1.97	2.54	1.25
	max factor diff *	9.2	58.4	17.3	46.3	39.9	47.2	54.1	13.6	46.2
SOIL	CalTOX	3.36E+04	2.44E+07	3.95E+05	1.56E+07	8.40E+07	7.43E+06	3.11E+06	6.46E+05	5.15E+10
	ChemCAN	8.83E+04	3.05E+07	2.92E+06	7.03E+07	1.47E+08	7.43E+06	4.48E+06	1.28E+06	9.40E+09
	HAZCHEM	8.67E+04	2.94E+07	2.90E+06	7.02E+07	1.46E+08	7.45E+06	4.54E+06	1.24E+06	6.20E+09
	SimpleBOX	1.60E+05	2.89E+07	4.18E+06	7.64E+07	1.64E+08	7.31E+06	6.15E+05	7.33E+05	6.81E+11
	coeff of var	0.47	0.09	0.48	0.39	0.23	0.01	0.57	0.31	1.42
	max factor diff *	4.8	1.3	10.6	4.9	2.0	1.02	7.4	2.0	109.9

* max factor diff: ratio between maximum and minimum

It is clear from these results that to be reliable, models must have fairly accurate expressions and parameters for describing inter-media transport. Potential for ambiguous statement of properties and concentrations is also apparent; input and output information must therefore be expressed in unambiguous terms.

The exercise was subsequently repeated in Denver for benzo(a)pyrene (B(a)P) and TCE, running simulations with simultaneous emissions of 1000 kg/h in air, water, and soil. The inconsistencies in the transfer and other parameters were accounted for by adjusting some parameters of the models in order to reach similar results. Some model-specific calculations had to be disabled (*e.g.*, the calculations for the different soils in HAZCHEM and SimpleBOX or the lower soil layers in CalTOX). The adjustments can be found in Table 3-5. It was found that ChemCAN, SimpleBOX, HAZCHEM, and CalTOX could give essentially the same results (within a few percent of the amounts for each compartment at steady-state). Figures 3-1, 3-2, 3-3, and 3-4 show the fluxes in kg/h among the compartments at steady-state (inter-media transfer), emissions in the system (kg/h), and the fluxes of chemical degraded or advected out of the system (kg/h). Amounts reached in each medium at steady-state (kg) are also reported. With effort, the models could probably be "tuned" to give exactly the same results to any desired precision, but this exercise was not deemed necessary since discrepancies of a few percent were judged to be trivial.

Based on the results of the round robin exercise, it was concluded that although superficial differences are apparent and the models do contain differences in parameter values, they are essentially similar and correctly represent the theory that they are intended to represent. However, it should be emphasized that

- the models produce essentially the same results (within a few percent) only if all relevant mechanisms and parameters are set to the same value.
- if some of the input parameters are left to the discretion of the modeler, different choices can be made yielding different results (Table 3-4). These differences in parameters, although understandable and for many scientific purposes insignificant, may be significant in a regulatory decision-making context.
- extension of the model definition with a list of values for mass transfer coefficients will eliminate most of the human error differences in model results, but certain structural differences in the models will result in discrepancies.

Considerable knowledge of the models is required to force the models to produce the same results, because some of the differences observed in the original comparison exercise were due to different mathematical descriptions of the same mechanisms. This is best done by the model developer. It is possible that different users could obtain different results from applying these models to the same data. This is especially likely if the users have a limited knowledge of environmental chemistry and modeling. Most of these errors would be due to ambiguities in specifying input data and representing output data, not to inherent differences in the models.

TABLE 3-5 *Sources of Differences Among HAZCHEM, ChemCAN, SimpleBOX, and CalTOX*

Difference	HAZCHEM	ChemCAN	SimpleBOX	CalTOX
Calculation basis mass balances	Fugacity basis (mass transfer based on D-values, derived from partial mass transfer coefficients and Z-values)	Fugacity and Aquivalence basis (mass transfer based on D-values, derived from partial mass transfer coefficients and Z-values)	Concentration basis (mass transfer based on overall mass transfer coefficients, derived from partial mass transfer coefficients and inter-media equilibrium constants) Adjustment: none; same mass balance principle	Fugacity and Aquivalence basis (mass transfer based on pseudo first order rate constants derived from partial mass transfer coefficients and Z-values)
Compartments	Bulk water compartment with equilibrium among sub phases	Bulk water compartment with equilibrium among sub phases	Biota and suspended matter as separate compartments Adjustments: equilibration time set to 0.001 h	Plant biomass compartment included, air depth = function of wind speed, soil includes three layers Adjustments: plant biomass set to zero, depth of air set to 1000 m, surface soil set to HAZCHEM soil depth.
Degradation in water	On bulk water basis	On bulk water basis	On (dissolved) water phase basis Adjustments: pseudo first-order rate constant water set to prescribed bulk rate constant / fraction dissolved	On bulk water basis
Phase volumes/densities	On solids basis	On solids basis	On wet volume basis Adjustments: volume fraction sediment and soil set equal to HAZCHEM; density solids set to 2400 kg/m^3 no other adjustments	On solids basis
Concentration biota in water	1 mg/l on wet weight basis	set to 1 mg/l on wet weight basis	1 mg/l on dry weight basis Adjustments: set to 0.1121 mg (dry solids)/l	Not used in mass balance
Bioconcentration	$BCF = 0.05\ K_{ow}$	$BCF = 0.05\ K_{ow}$	$BCF = 0.05\ K_{ow}$ / 1.07 (density biota 1070 kg/m3) Adjustments: set to HAZCHEM value	$BCF = 0.05\ K_{ow}$
Solids-water partitioning	$K_p = f_{oc}\ 0.41\ K_{ow}$	$K_p = f_{oc}\ 0.41\ K_{ow}$	$K_p = f_{oc}\ 1.\ K_{ow}$ Adjustments: set to HAZCHEM value	K_p=foc 0.48 K_{ow} Adjustments: set to HAZCHEM value

TABLE 3-5 — *Continued*

Difference	HAZCHEM	ChemCAN	SimpleBOX	CalTOX
Aerosol-gas partitioning	$Z_{aerosol}/Z_{gas}$ = function of P^S_L;	$Z_{aerosol}/Z_{gas}$ = function of P^S_L;	Fraction associated with aerosol = function of vapor pressure; constant = 10^{-4} Adjustments: fraction associated with aerosol derived as in HAZCHEM	Fraction associated with aerosol = function of vapor pressure and melting point Adjustments: derive as in HAZCHEM
Dry deposition aerosol	10 m(air)/h	set to 10 m (air)/h = 1.2E-6 m(solids)/h	0.1 cm(air)/s = 3.6 m/h Adjustments: set to HAZCHEM value	2.1 m(air)/h Adjustments: set to 10 m(air)/h
Volatilization / absorption water	k_{air} = 5 m/h; k_{water} = 0.05 m/h	set to k_{air} = 5 m/h; k_{water} = 0.05 m/h	k_{air} and k_{water} = function of wind speed Adjustments: set to HAZCHEM value	k_{air} and k_{water} = function of wind speed and water current Adjustments: set to HAZCHEM value
Volatilization / absorption soil	$k_{soil\ air}$ includes capillary transport	$k_{soil\ air}$ includes capillary transport	$k_{soil\ air}$ gas phase diffusion only Adjustments: set to HAZCHEM value	$k_{soil\ air}$ = function of effective diffusivity and boundary layer thickness; includes capillary transport Adjustments: Set mass transfer coefficients to HAZCHEM value
Diffusion / Sedimentation / Resuspension / Burial	Sedimentation = 8.76 mm(solid)/y; Net sedimentation = burial = 1.75 mm(solid)/y; Resuspension = sedimentation - net sedimentation	Sedimentation set to 10.5mm(solids)/y Sed. resuspension set to 6.57mm(solids)/y Sediment burial set to 1.75mm(solids)/y	Sedimentation = function of suspended matter concentration and settling velocity; Net sedimentation = computed from solids mass balance; Resuspension = sedimentation - net sedimentation Adjustment: sedimentation set to 43.8 (= 5x8.76) mm(wet sediment)/y; burial set to 8.75 (=5x1.75) mm(wet sediment)/y	Sediment-water diffusive exchange includes water side mass transfer, Sedimentation set to 8.5 mm(solid)/y, Net sedimentation = burial = 1.46 mm (solid)/y, Resuspension = computed from solids mass balance Adjustments: diffusive mass transfer derived as in HAZCHEM, net sedim. = 8.76 mm(solid)/y, burial = 1.75 mm(solid)/y, resuspension = sediment - burial
Soil run-off	Based on erosion of 0.1 mm(solids)/y	Based on erosion of 0.1 mm(solids)/y	Based on erosion of 0.1 mm (soil)/y Adjustments: erosion set to 0.175 mm (soil)/y	Based on erosion of 0.05 mm(solid)/yr Adjustments: set erosion to HAZCHEM value
Soil leaching / Diffusive transfer /Resuspension	No advective loss to sub-surface soil/groundwater	No advective loss to sub-surface soil/groundwater	Loss to groundwater by percolation Adjustment: infiltration set to zero	Loss to deeper soil layers by infiltration and diffusion, loss to air by particle resuspension Adjustments: infiltration set to zero, diffusive migration to root soil set to zero, resuspension set to zero

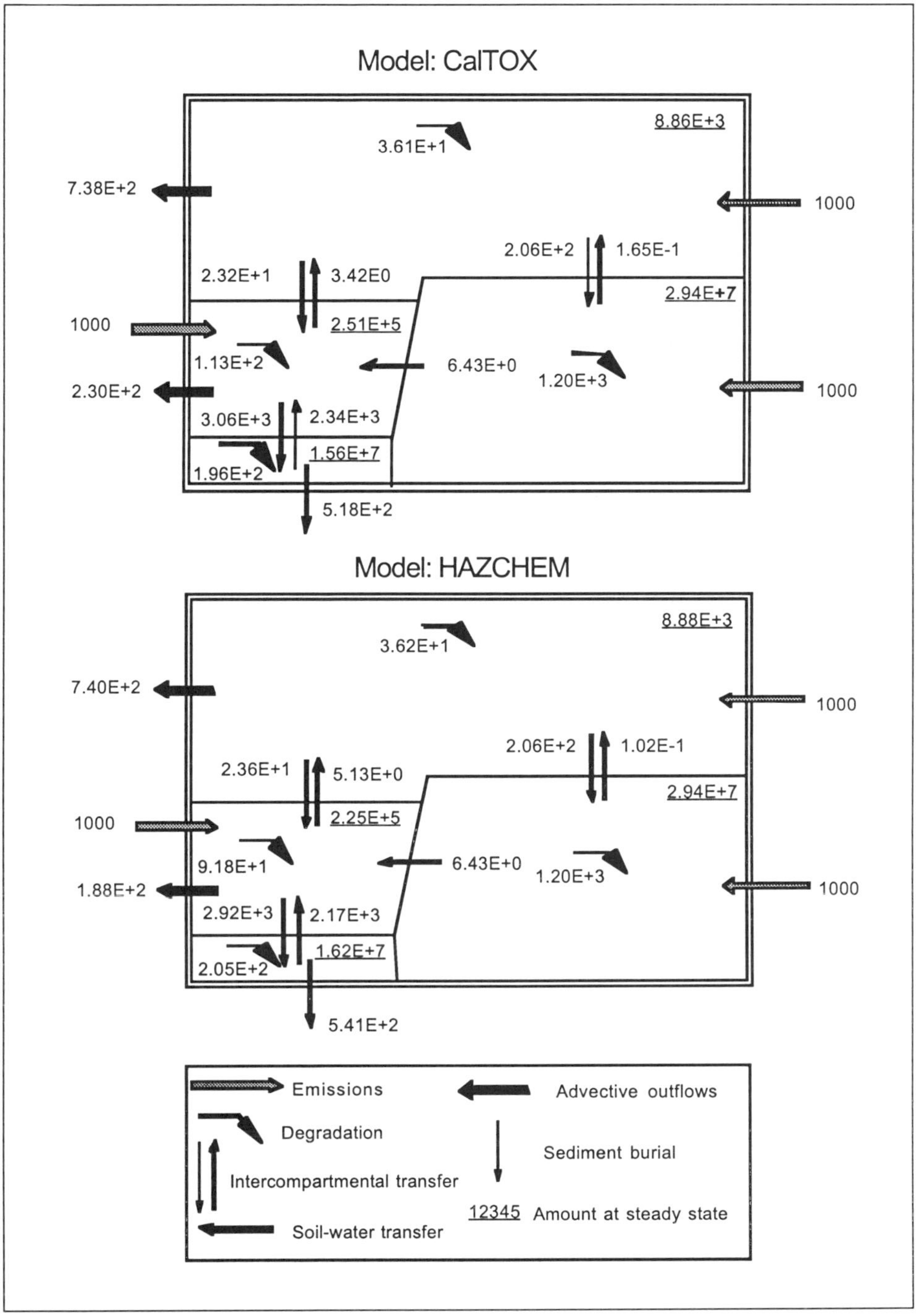

Fig. 3-1. Results of the simulations for B(a)P for CalTOX and HAZCHEM models (All rates are in kg/h; amounts are at steady-state in kg.)

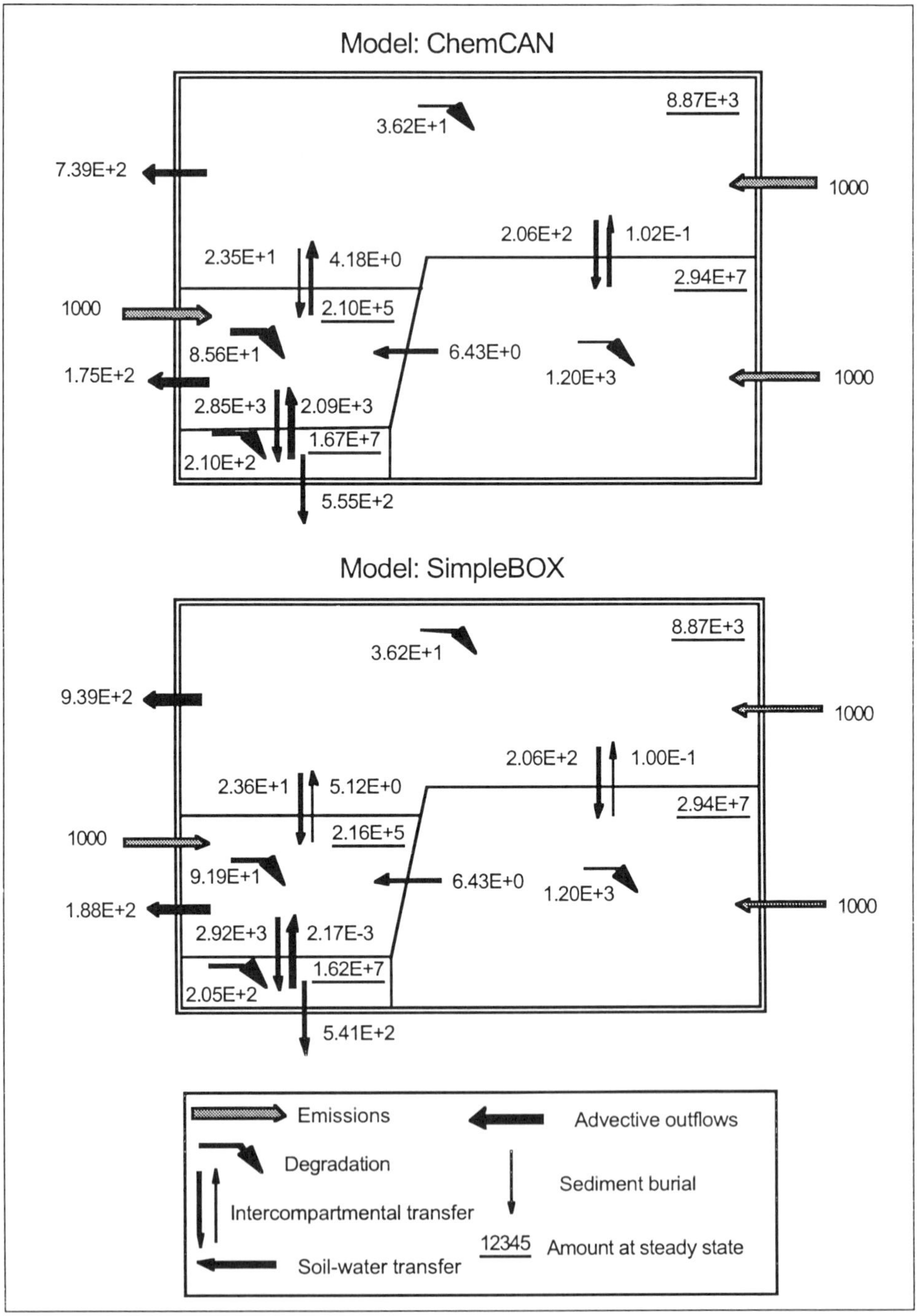

FIG. 3-2. Results of the simulations for B(a)P for ChemCAN and SimpleBOX models (All rates are in kg/h; amounts are at steady-state in kg.)

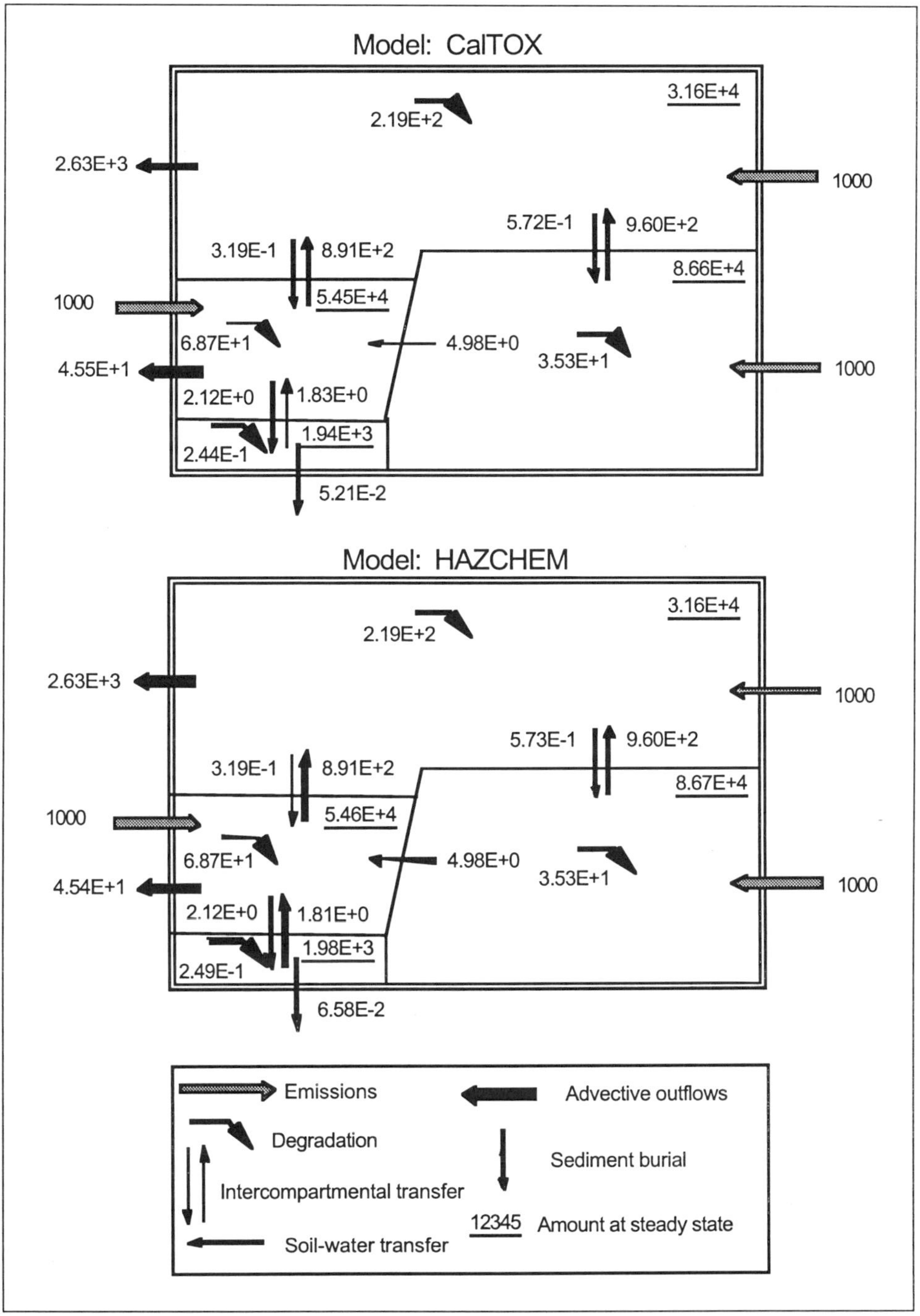

Fig. 3-3. Results of the simulations for TCE for CalTOX and HAZCHEM models (All rates are in kg/h; amounts are at steady-state in kg.)

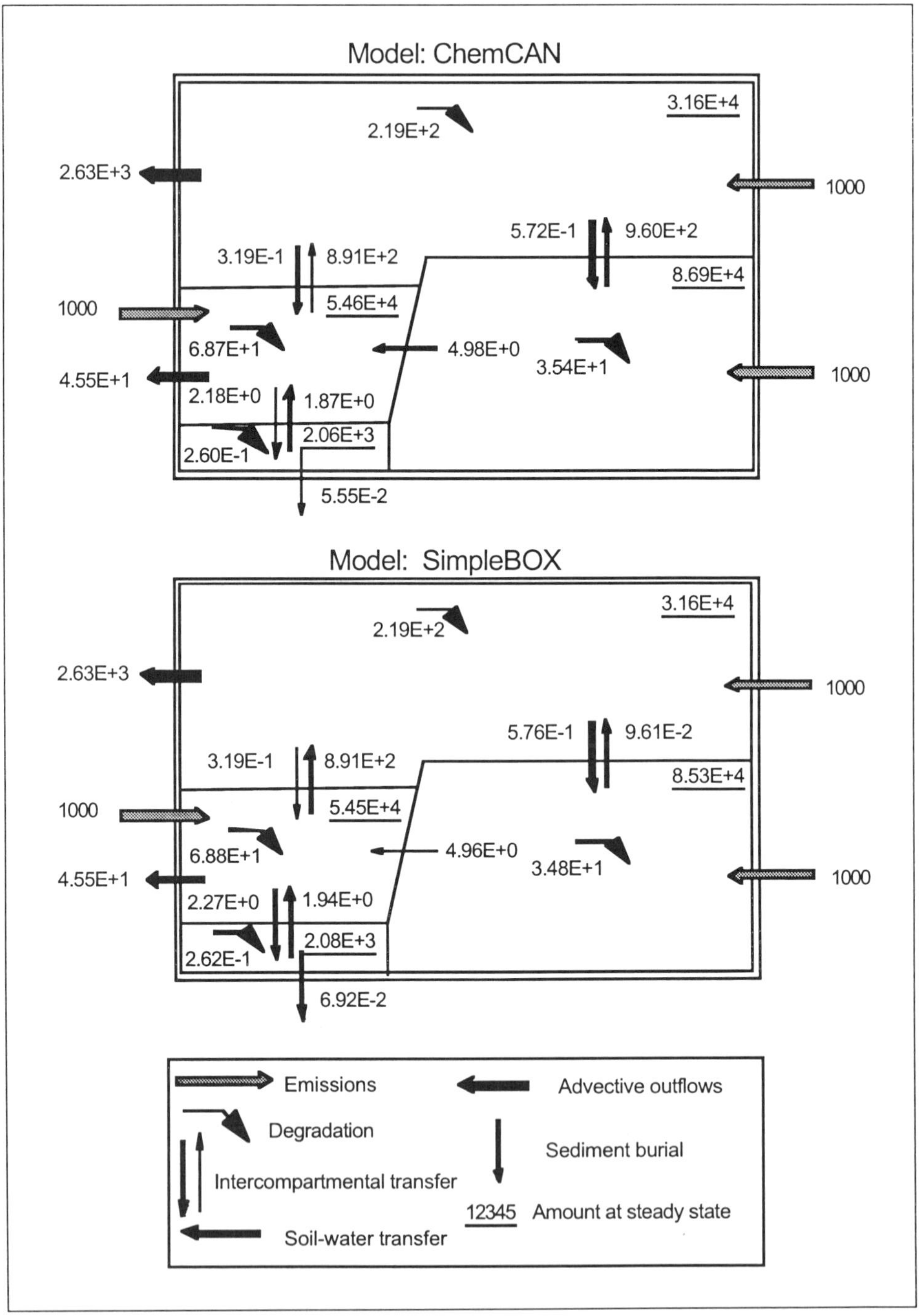

FIG. 3-4. Results of the simulations for TCE for ChemCAN and SimpleBOX models (All rates are in kg/h; amounts are at steady-state in kg.)

The modelers were very encouraged by the consistency among the models. They found that the round robin exercise was of immense value for promoting collaboration, and was an invaluable vehicle for cooperatively improving model performance. It is hoped that this type of exercise will be repeated periodically. Furthermore, a great deal was learned about the conduct of such exercises from this pioneering effort.

3.4 Ensuring Model Harmonization

Given the results of the model comparison and the concerns relating to the potential for errors even among experienced users, most, if not all, participants concluded that it is preferable that a single model be used for a given regulatory application. Errors in both data interpretation and differences in model results will be especially problematic in regulatory situations where the effort and results of the modeling need to be shared among regulatory bodies. Examples are the ranking and priority setting for a set of chemicals and the sharing of fate predictions in international agencies such as OECD. The user group felt that a single agreed model was a way to increase consistency. Furthermore, a single model is preferable in regulatory situations where the input data sets and the format of the output are well defined. Although development of a single model could reduce errors and would reduce differences among simulations, the choice of a single model does not ensure that this model will have the required precision, will be consistent with the results of the models presented here, and will faithfully express the theory that it is intended to represent.

This issue was the subject of considerable discussion at the workshop, especially among the model developers who were concerned that if only one model were defined and used there could be inhibition of future developments including innovative approaches.

To ensure precision among models used by different agencies and for different applications, the participants decided that it would be appropriate to specify that any model used in regulatory applications meet specific performance criteria to facilitate model comparison. These performance criteria would be defined as the ability to generate the same chemical fate results (within a difference of a few percent) for a number of chemicals discharged under specified scenarios into a defined environment. This approach is similar in principle to specifying that an analytical method or instrument be capable of generating a specific result, without providing the details on how it is to be done. Specifying this capability rather than requiring the use of a single model recognizes that the applications of these models, and thus the input and output interfaces, will probably be specific to the intended use. The model developer would be responsible for ensuring that the model meets the performance criteria. After this is shown, the input and output interface could be customized by the model developer to take advantage of any prior knowledge about the input data and the portions of the output data required for that application. These customized interfaces on a "precise" model could be used to reduce the potential for misuse by non-expert users. The participants agreed that even when the model is shown to be "precise" there still needs to be an emphasis on ensuring quality control on model results and their use. The initial focus is likely to be on establishing performance criteria relating to model precision, *i.e.*, ensuring that models give the same result. Later, when validation data sets are available, the criteria can be extended to include accuracy, *i.e.*, the ability of the model to predict field data reliably. There may therefore be separate criteria for precision and accuracy.

The workshop participants recognized that performance criteria need not, and indeed should not, remain static. There should be provision for periodic review and updating of criteria every two to three years. They also recognized that there should be provision for users to adopt different approaches when circumstances, such as advancement in understanding on how to represent partitioning and transport and transformation processes, justify such a deviation. This is especially true for vegetation, soil, and sediment compartments where further research is ongoing and is clearly needed before a consensus can be developed on how best to model these compartments.

Since some of the discrepancies in the round robin exercise were due to the QSARs used to calculate secondary model parameters, the workshop participants also recommended that there be explicit recognition in the documentation of the QSARs that are used in a model and a statement of the degree of uncertainty. Whenever possible, the most universally accepted QSAR should be used, but it is expected that different users will choose different QSARs. Therefore to ensure greater flexibility, the participants recommended that the user be able to override the QSAR predictions and directly enter the secondary parameters. In addition to providing flexibility to the user, this capability will also facilitate model comparison, as is required when ensuring compatibility with specified performance criteria.

In addition, the participants recommended that a set of criteria be developed and implemented for documenting any model that is to be used in a regulatory application. Some of the mechanisms for conveying information on the models and their limitations are described in Section 4 of this report. All of these activities should help meet the need for a high degree of model precision and to provide the user with the information necessary to understand the realm within which the models can be applied appropriately.

4: Guidance on the Application of Multi-Media Models

Given the results of the model comparison and the needs of the users, a number of recommendations were developed by the workshop participants regarding desirable features of model structure and how to best describe model limitations and capabilities. These issues are presented in some cases as answers to specific questions.

4.1 Environmental Compartments and Processes

Given that the amount of partitioning and reactivity data required increases with the number of environmental compartments treated, and that the data available for most of the applications described previously are likely to be limited and of uncertain quality, the workshop participants focused on defining the minimum compartments and processes that should be included to meet the goals of most applications. The consensus was that at a minimum the model should include the compartments and gain and loss processes specified in Table 4-1. Three land uses were recommended to allow for the inclusion of a reasonable range of soil types within a region. Transport to the environment beyond the area modeled should be specified, including air movement, water movement, and transport to the stratosphere.

TABLE 4-1 *Minimum Multi-Media Model Components and Processes*

Compartments	Gains	Losses	End Points
Air with aerosols	Emission into air Inflow from neighboring region Volatilization from soil and water	Horizontal and vertical advection Wet and dry deposition Degradation reactions Transport to stratosphere	Concentration in air and on aerosols Advective loss Reactive loss
Three surface soils representing different land uses	Emission to soil Wet and dry deposition	Volatilization to air Degradation reactions	Concentration on solids and water Reactive loss
Surface water with suspended particles and biota	Emission into water Inflow from neighboring region Wet and dry deposition Desorption from sediment	Volatilization to air Advection out of region Degradation reaction Sorption to sediment	Concentration in water, on suspended solids and biota Advective losses Reactive losses
Sediment	Sorption from water	Desorption to water Degradation reactions	Concentration on solids and interstitial water Reactive losses

Figure 1-1 was used earlier to illustrate the key processes which control the fate of chemicals in a simple four-compartment environment. Although not all processes are important for all chemicals, there are at least some chemicals for which each process proves to be significant. As is discussed later, there are differences in opinion about how some of these processes should be expressed and parameterized, but the key requirement is that they be included using realistic expressions with reasonable parameter values.

As a consequence of the way that soil is modeled in most of the multi-media models, they cannot address groundwater contamination from either buried waste or from movement of chemicals through soil to the groundwater. The model developers therefore recommended that, with the exception of the CalTOX model which treats chemical migration through soil in more detail, multi-media models not be used to predict or infer realistic groundwater concentrations because groundwater flows and media attenuation conditions are so variable and site-specific. The model developers suggested that a soil pore water concentration could be reported as a maximum source term to groundwater.

4.2 Food Chains and Indirect Exposure

Quantitative estimates of bioaccumulation at higher trophic levels are needed to accurately assess indirect exposures. The participants agreed that inclusion of the secondary and tertiary consumers of aquatic and terrestrial food webs as additional compartments in multi-media models was not required to predict the overall mass distribution of a chemical. This is due to the relatively small biomass and consequent small quantity of chemical that would be contained in such compartments. The one exception to this is vegetation which may significantly affect the chemical's mass balance. Therefore, it is recommended that the trophic transfer of contaminant through food chains not be included directly in multi-media models. Rather, it is recommended that the output from multi-media models (*i.e.*, concentrations in primary abiotic compartments) be used as input to separate food chain and indirect exposure assessments, especially when the focus is on regional fate and exposure to a chemical with numerous dispersed releases. In addition to providing estimates of indirect exposure from regional contamination, these coupled multi-media and indirect exposure models can be used to focus field monitoring programs on the most relevant dietary pathways.

The accuracy of these coupled multi-media and indirect exposure models varies with the chemical being considered. Although current aquatic food-chain-based bioaccumulation models yield predictions within an order of magnitude for poorly metabolized non-ionic organic chemicals with log K_{OW} values less than 5, for more hydrophobic compounds (log $K_{OW} > 5$), the predictions are expected to be no better than plus or minus an order of magnitude without site-specific data on how hydrophobic chemicals partition to phytoplankton/seston at the base of the aquatic foodweb. Most current multi-media models characterize contaminant partitioning in the water column as a simple equilibrium process that is correlated to the chemical's octanol-water partition coefficient. However, field data are often inconsistent with this assumption (Baker *et al.*, 1991; Swackhamer and Skoglund, 1993). Therefore, if more accurate predictions are required, a more accurate description of contaminant partitioning in the water column must be incorporated into the multi-media models, or a site-specific model may be required.

4.3 Chemical Property Data Requirements

In order to predict the distribution and fate of a chemical in the environment, a multi-media model must specify the chemical's basic properties: partitioning among the relevant environmental media, rate of transformation/reactivity in the relevant media, and rate of discharge into each environmental medium. Partitioning can be (1) specified in terms of the measured partition coefficients and/or mass transfer coefficients or (2) calculated as a secondary parameter from the chemical's properties such as solubility, vapor pressure, and K_{OW}. Reactivity is usually provided as a half-life for each transformation process, *i.e.*, hydrolysis, biodegradation, and photolysis. Discharge estimates are represented as loading rates into specified media for the region of interest. As discussed earlier, the amount of partitioning and reactivity data required increases as the number of compartments increases. Thus, there is always a trade off between the amount of data available and the number of compartments that are included in the model. For applications where discharge, partitioning, and reactivity data are likely to be the most limited, the participants recommended that the simplest model with the least number of compartments be used because the results from this model will have greater credibility than those of a more complex model that has inadequate input data.

As noted in the discussion of the application areas, the amount and type of data available on a chemical of interest will vary depending on the chemical and the required accuracy of the final results. The user group requested that whenever possible the multi-media models use the available data, minimize the generation of additional data, and if possible, include methods for estimating data to fill any data gaps. Without a more detailed description of what data are expected to be available for each application, it is difficult to know how limited the available data will be. QSARs or QSPRs are available for estimating the partition coefficients from other chemical data or even in some cases from the chemical's structure (*e.g.*, Lyman *et al.*, 1982). However, QSARs exist for only a few classes of chemicals (*i.e.*, nonpolar, hydrophobic organic chemicals, and some polar chemicals). Their ability to predict accurately the partition coefficients for other classes of chemicals is the subject of ongoing research. Fewer QSARs are available for predicting half-lives or rates of transformation, and they are to a large extent limited to homologous chemicals. Reactivity data, usually in the form of half-lives, are critically important parameters in most assessments. As indicated previously, the assessments will in many cases be iterative. Thus, if the initial assessment suggests that one of the estimated values is very influential, then research could be focused on generating high quality data for this parameter. In cases such as risk assessment for notification and/or registration of new chemicals, where the available data are defined in the regulations and will increase with increasing volume of production (EU and CEPA), then the models need to be flexible and allow for parameters to be estimated or entered directly. Whether QSARs or QSPRs will be included in the input interface to the multi-media models or handled separately will depend on the specific application.

Beyond the basic partitioning and transformation rate data, other data will be required depending on the objective of the application. When accurate predictions of environmental concentrations are the objective of the application, data on the discharge quantity and media into which the chemical is discharged are critical. It cannot be over-emphasized that it is input data in the form of emissions and advective inflows which "drive" the models because, to a first approximation, concentrations are proportional to these input quantities. Yet, these are often the least well-known input data in the entire modeling exercise. When the effect of temperature on the chemical's distribution is part of the focus of the assessment, then thermochemical data are needed to adjust rate constants that are affected by the temperature variation. If concentrations in biota are important, then bioenergetic information, *e.g.*, metabolic rates in animals and plants, may be required for use in supplementing food chain models.

4.4 Chemical Domain or Chemical Classification

Because model formulations assume certain partitioning and transformation processes and, in some cases, include QSPRs for estimating secondary parameters such as partition coefficients, the multi-media model may be applicable to only certain classes of chemicals. Historically, multi-media models were developed for predicting the fate of nonpolar, hydrophobic organic chemicals because these chemicals were of environmental concern. Furthermore, there were extensive data available on their physical-chemical properties (Mackay *et al.*, 1992) and QSARs for estimating data that were not available (Lyman *et al*, 1982).

To meet the need for better definition of the appropriate chemical domain for each model, the workshop participants recommended that in all model documentation there be an explicit statement as to which classes of chemicals the model can, and cannot, be applied. This will help to prevent or reduce misapplication. Proper classification of the chemicals to be evaluated will also identify the appropriate data requirements. Obviously, there is little value in measuring the partitioning to or from, or reactivity in, a medium in which the chemical is very sparingly or negligibly soluble. Because there is no single agreed scheme for classifying chemicals, the workshop participants suggested a classification scheme that could be used to indicate which chemicals are suitable for model application. This scheme may be used in addition to the common chemical classification schemes, *e.g.*, nonpolar, polar, volatile. Use of this scheme or similar ones will help to standardize and increase communication on the applicable chemical domain of multi-media models.

The proposed classification scheme (Mackay *et al.*, 1995a) uses the various types of partitioning among the major environmental media as well as speciation as its basis. This results in six groups or classes:

- volatile or partially volatile organic chemicals (*e.g.*, chloroform, polychlorinated byphenyls [PCBs], polynuclear aromatic hydrocarbons [PAHs])
- involatile chemicals (*e.g.*, metals, surfactants)

- insoluble chemicals (*e.g.*, chloroparaffins, silicones, waxes)
- chemicals which are both involatile and insoluble (*e.g.*, dyes, polymers)
- certain speciating chemicals (*e.g.*, mercury and phenols)
- fast-reacting chemicals (*e.g.*, hydroquinone)

It is recommended that the terms *involatile* and *insoluble* imply absolute absence from a phase. In reality all chemicals are slightly soluble in all media, thus these terms should be interpreted to imply absent for all practical purposes.

Most of the current multi-media models, as exemplified by the models evaluated in the workshop, cannot adequately treat all these classes of chemicals, especially speciating chemicals and to a lesser extent surface-active chemicals. The workshop participants also recommended that extreme care be taken when modeling the fate of chemicals of uncertain identity, chemicals that occur in mixtures, and metabolites and degradation products. None of the models are capable of handling chemicals as pure or mixed organic phases, such as non-aqueous liquids or solids frequently encountered in groundwater contamination situations. The model developers also recognized that site-specific models, in addition to qualitative multi-media assessment, may be required to evaluate pesticides because these chemicals are biologically active and are applied directly to the environment in relatively small areas. However, rather than make broad statements about multi-media model applicability to classes of chemicals, the participants felt that it was more appropriate to demand a definition of the chemical domain which can be addressed by each model as currently perceived by the model developer. It is recommended that users look for or demand a statement of these limitations and avoid the dangers of assessing chemicals with inappropriate models.

The workshop participants were cognizant of the pressures to apply a model when there is no suitable alternative; however, they felt that these misapplications would eventually not only discredit the models but also the users. It was concluded that it is better to require model developers to construct suitable models. The developers felt that the theory existed from which models could be developed for all the classes of chemicals but that in some cases the models could become quite complex, as would be the case for speciating chemicals.

4.5 Environmental Characteristics

The variation in the values of the parameters used to describe the environment accounted for much of the variability found when model results were compared. An effort to reduce this variability and to facilitate comparison of multi-media model results, but still allow for country- or region-specific environments, would be beneficial. One option is to use default data sets for representative environments as suggested by Mackay (1991) and Mackay *et al.* (1992) and to distribute these with the multi-media models. To facilitate international communication and sharing of understanding, the default data set used would be specified (*e.g.*, Mackay *et al.*, 1992) when model results are discussed. Another approach to meeting this goal, while not constraining the types of regions and regional characteristics that can be modeled, would be to couple scenarios and transport parameters in such a way as to guide the user in the selection of the best values. The approach suggested by the model developers follows.

Within a region simulated by these multi-media models, there can be various mixes of environmental types or Uniform Geographical Units (UGU) with very different characteristics. A *Uniform Geographical Unit* is defined here as an area of geographic surface, fairly uniform in its physical and biological characteristics, such that it can be represented by a specific set of transport and characterization parameters. For example, a model for the wheat-growing region in France would have very different environmental characteristics from an estuarine region in the Netherlands. To increase the ability to model this wide range of geographical units, the model developers recommended that several generic environmental characteristic data sets be defined which represent important UGU. Those units that have been tentatively identified as being sufficiently different to merit unique environmental characterization data sets (Table 4-2) are divided into two basic types: aquatic and terrestrial. Further subunits are possible.

These generic data sets could also be combined to define specific regions. For example, once a specific region and physical scale are chosen for the model application, a geographic data base, such as a Geographical Information System (GIS), could be used to determine the types and relative areas of each UGU. Using this information, an algorithm can be developed to compute the transport coefficients and environmental characteristic values representative of the region of interest. Alternatively, expert judgment based on other information can be used to assign parameters based on the most suitable unit and the range in parameter values among units. Examples of special regions that could be created from combinations of the UGU in Table 4-2 are an island (UK), a peninsula (Italy), and a lakes region (Canada-U.S. Great Lakes). The advantage to this approach is that communication of the environmental characteristics used in a specific assessment would be reduced to communicating the UGU used and the algorithm used for computing the representative characteristics for this region.

TABLE 4-2 *Representative Uniform Geographical Units*

Aquatic	Terrestrial
Rivers/Streams	Forest
Lakes	Uplands
Seas	Grasslands
Oceans	Lowlands
Estuaries	Desert
Continental Shelf	Agricultural Lands
	Urban
	Industrial
	Mixed Land Use

Of all the environmental parameters required to characterize a UGU or region, the transport coefficients between media are among the most important. These coefficients are defined as velocity constants or mass transfer coefficients used to deduce a flux from a concentration difference between two media (*i.e.*, flux = $k\Delta C$). Based on the two-resistance theory, ideally there are usually two coefficients, one located on either side of the various medium interfaces, *i.e.*, air-soil, air-water, water-sediment, *etc.* (Thibodeaux, 1979). There is significant variation in these coefficients in space and time between various regions. Those at the air-water interface are best known, but a factor of 10 variation still exists. For example, for windy conditions at sea the air-water transport coefficient value may be 10 cm/h (O_2 basis), while a coefficient value of 2 cm/h may apply for a Louisiana bayou protected from the wind by a tree-lined bank. By employing the UGU, these variations can be accounted for in the environmental data set.

Although all transport coefficients can be expressed as velocity constants, there are fundamental differences between the fluid-phase coefficients for water and air as compared to the solid-phase coefficients for soil and sediments. The turbulent mixing of the fluids near environmental interfaces provides a theoretically consistent justification for using single parameter, mass transfer coefficients for these transport parameters. In the case of soil and sediment, these interface regions are not usually well mixed. In this case, the theoretical approach involves specifying a diffusive transport parameter (D) and an effective depth scale (d) within the soil or sediment. The velocity constant can then be regarded as D/d.

It should be possible to compile a data base of reasonable values for these transfer coefficients. The compilation of these coefficients for a wide variety of environments is important to the user of environmental fate models for several reasons. First, these parameters can be used to characterize and define UGU. These UGU can be used in applications where generic environments representing typical regional units (*e.g.*, a lake region, rivers, agricultural land) would be most appropriate. An example is a regulatory situation in which information on the fate of a substance in a specific type of environment is to be communicated internationally. Second, these parameter values can be used to educate model users regarding the natural range in variability and the importance of a particular selection of transport coefficients. Finally, these compilations will assist in the development of environmental definitions for UGU and with direct research into determining transport coefficients.

4.6 Appropriate Scales in Space and Time

An important issue in the application of multi-media models is the effect of variation in the spatial and temporal scales of the model. Variations in these scales can greatly affect the average concentration in an environmental compartment. This is especially important when the assessment focuses on a single point source of chemical, which generates high concentrations near the discharge point that decay with distance. This is illustrated by the following example. If for a given discharge, a model treating a 1 km^2 area predicted a concentration of 1 mg/L, then for the same discharge but with the model now representing a 10 km^2 area, the concentration may be 0.1 mg/L, and for a 100 km^2 area the concentration may be 0.01 mg/L. This implies that by selecting the appropriate size of the area modeled, any desired concentration can be achieved. The model is correct in expressing the average concentration over the region, but this can be misinterpreted as a local concentration. To avoid such problems, the model developers recommended that multi-media models of the type discussed here not be used to predict concentrations from point sources, that they use emissions rates that are appropriate for the region modeled, and that these emissions be dispersed within the region. For example, multi-media models should not be used when the source term is derived from a massive spill of chemicals into the soil or into the underlying geologic strata.

There is a concern that multi-media models could be applied inappropriately to model the fate of chemicals at too small a scale. At the smaller scales, the key assumption of uniform mixing of the chemical in the environmental media becomes invalid. For chemicals discharged to the atmosphere, the smallest recommended scale of the model would be equivalent to the distance at which a dispersing plume reaches the mixed layer depth. This will generally be 5 to 10 km. Thus the model developers recommended that multi-media box models not be used to quantitatively predict concentrations of chemicals in areas less than a few hundred km^2. For quantitative predictions at smaller scales, other approaches and models should be used, such as air dispersion or river dilution models (*e.g.*, Rapaport, 1990). Multi-media models can still be used, however, to gain qualitative understanding of the processes affecting a chemical's fate at these smaller scales.

Most of the available multi-media models are primarily steady-state in nature. They should therefore be applied to time scales that are long relative to the *characteristic time* of the model system. The characteristic time is the mean residence time of a chemical molecule in all the model compartments. This time can be calculated under steady-state conditions as the total system inventory divided by the total input or loss to the system. Further, the emissions into the environment should be relatively constant over this residence time. For chemicals whose emissions are highly variable, such as pesticides, these steady-state multi-media models should only be used to obtain a qualitative understanding of the influence of transport and transformation processes on their fate. However, Level IV model may be applicable.

4.7 Sensitivity and Uncertainty Analyses

The workshop participants recommended that some form of sensitivity and uncertainty analyses (*e.g.*, McCuen, 1973; Iman and Helton, 1988; Boesten, 1991) be included in the multi-media model assessments. In order to address sensitivity and uncertainty, one can think of a model as producing an output Y, such as concentration, that is a function of several input variables, X_i, and in the case of unsteady-state models with time t

$$Y = f(X_1, X_2, X_3, \ldots X_k, t)$$

4.7.1 Sensitivity Analysis

Sensitivity analysis is defined as the study of how variations in the different model input parameters, X, around their nominal values affect the output, Y. For example, if a log K_{OW} value with a variability of ±0.5 log units is used as input, the sensitivity analysis will determine the corresponding variability in the output of the model by performing simulations using log K_{OW} values over this range. Similar analyses could be performed for all the input parameters. At the end of these analyses, a ranking of the influence of each of the input parameters on the output using statistical and/or graphical methods (Devillers *et al.*, 1991) could be used to identify the critical model input(s). As discussed in Section 2.5, these results can be used to guide the need for further refinement of input parameters.

4.7.2 Uncertainty Analysis

Uncertainty analysis is defined as the study of all the uncertain aspects of a model and their influence on the model output, Y. Therefore, it evaluates not only the effect of uncertainty in the input data, X, on the output, Y, but also the effect of assumptions made by the modeler during conceptualization, the reduction, and the mathematical representation of the region. There are four recognized sources of uncertainty (also sometimes called *variance*) in the data that are used in models, and a fifth source lies in the inherent structure of the model:

- deterministic variance (*e.g.*, experimental or predictive error) in quantities such as solubility which can be reduced by better measurement
- stochastic variance (*e.g.*, variation in nature) resulting from changing environmental conditions (*e.g.*, rain rate) which are inherent and cannot be controlled
- variance in reaction rates or half-lives which are both stochastic (because of differences in the environment) and deterministic (because of uncertainty in chemical properties), *e.g.*, volatilization
- lifestyle variances such as age, ingestion rates, and food preferences which are used in exposure calculations
- model uncertainty including its structure, the detail of equations used, the scenarios to which it is applied, and the extent of extrapolation

It is important to recognize that multi-media models appear to be most sensitive to variability in release data and reaction data (*e.g.*, half-lives). Therefore, to reduce uncertainty in model output, efforts are best focused on determining these parameters with the greatest accuracy. In addition, there are always the unknowns around the model formulation, especially the simplification of environmental processes. It is very difficult to quantify this source of uncertainty, since the possibility always exists that the modeler applies equations which are inherently inappropriate. Of course, no modeler will knowingly do this, but such a possibility is always present and by its very nature is not easily detected. Perhaps the best approach to detecting such problems is to compare the results of independently derived models applied to the same task.

In the process of representing chemical behavior within a landscape unit, the structure of mathematical models employed can, in some cases, be a key source of uncertainty. This is due to the fact that models are, by their nature, only a simplified representation of a real-world system. Significant approximations are often an inherent part of the assumptions upon which a model is built. For small, well-calibrated physical models, model uncertainty can and should be significantly less than input variance. However, for large, complex models that have many sub-models as components, model uncertainty can easily surpass parameter uncertainty as the most significant contribution to the variance in model outputs.

The main sources of model uncertainties include model structure, model detail, validation, extrapolation, model resolution, model boundaries, and model scenarios. Model structure refers to the main components of the model, for example, the compartments. Failure to include all relevant compartments can reduce model reliability. Model detail refers to the complexity of feedback and time rate of change in a model, *i.e.*, steady-state versus dynamic or linear-independent versus non-linear coupled compartments. Approximations of these details, such as substitution of a linear model to represent potentially non-linear processes, can limit the domain for which the model is valid. The limited spatial and temporal resolution of many multi-media models is an approximation that is likely to introduce some potentially significant uncertainties. The extent to which a model has been validated should increase its expected reliability. Extrapolating models beyond the range of parameter values for which they have been calibrated increases the uncertainty associated with the model results. Similarly, any model has limited boundaries in terms of space, time, chemical species, temperature ranges, *etc.* The selection of model boundaries is a type of simplification that can lead to significant uncertainties if the model is applied to a domain outside of these boundaries. Scenarios are sets of assumptions about the nature of the problem to be analyzed. To the extent that built-in model scenarios, *i.e.*, continuous releases, fail to apply to the situation for which the model is applied, a loss of reliability can be expected.

4.8 User Qualifications

Frequently, model developers are asked how experienced the user must be to obtain reliable results, especially given the wide diversity in background of the potential users, which ranges from B.Sc. to Ph.D. and from training in general science to environmental chemistry. The group was of the opinion that a sound knowledge of environmental chemistry, including inter-media transport phenomena, is essential for independent use. With more user-friendly models being developed, it is likely that models will be increasingly used by less experienced individuals so that the danger of misuse will increase. Inclusion of "checking" or "fail-safe" or "error detecting" systems may help, but it is likely that a fully safe and friendly program will be 25% model, 75% "help" systems. The model developer group was not confident that a system could be devised which would remove all opportunities for misuse.

The model developers recommended that the major effort be placed on ensuring quality control on the model results and subsequent use, rather than on defining minimum requirements for the users. Inexperienced users may find benefit from using models they do not fully understand. If an agency plans to use model results to support regulatory decisions, it must ensure that input data, model, and results are all reliable, and that interpretation is appropriate. They also recommended that when a specific model is developed for repetitively producing a specific output and for addressing a very specific regulatory question, model codes should be produced that allow relatively inexperienced users to avoid misuse.

4.9 Vegetation Compartment Treatment

Plants live by interacting with soil (uptake of water and minerals) and the atmosphere (uptake of CO_2, O_2, and sunlight) and thus accumulate chemicals from both air and soil. Recently, Simonich and Hites (1994) demonstrated that vegetation could play a very important role in determining the fate of certain hydrophobic organic chemicals. There is increasing acceptance of the view that inclusion of vegetation biomass is essential for determining the appropriate environmental mass balance of these chemicals. For example, monitoring of plant leaves and pine needles indicates that up to 10% of B(a)P released into air is associated with vegetation, depending on the biomass quantity and the methods by which the concentration is estimated. In addition to being important in the mass balance of certain hydrophobic organic chemicals, accumulation of chemicals in plants may be a human and ecotoxicological health concern due to direct consumption of vegetation or indirect consumption via entry into food chains. Concentrations in vegetation have been found to be of the same order of magnitude as those of aquatic organisms or soil (Trapp and McFarlane, 1995).

The model developers discussed how the uptake of chemicals by the vegetation should be represented in multi-media models. The following approaches were recommended:

- **Roots** of most vegetation, with the exception of a few root crops such as potatoes, are not harvested. Thus, when the role of roots in chemical fate and mass balance is evaluated, they can be included in the models as an increase in the soil organic carbon above that of the soil matrix. When the concentration of chemical in the root is the goal of the evaluation, then the uptake from soil into roots can be calculated using a root-concentration-factor (RCF) (Briggs *et al.*, 1982).
- **Aerial plant parts** (except fruit) may be best treated as one compartment. When the concentrations in foliage are sought, the mass balance equation

for this compartment should take into account (1) uptake by the roots and the translocation from the roots to the aerial plant parts with the transpiration stream, (2) gaseous exchange with the air (uptake and volatilization), (3) wet and dry deposition onto foliage, and (4) photo- and/or bio-metabolism. An example is the work of Bacci *et al.* (1990).

The model developers suggested that both roots and aerial vegetation be characterized in the UGU or region to be modeled by a standard (median) biomass and lipid content, which would depend on the type of vegetation included in the region of interest. This approach is analogous to that used to represent aquatic organisms. The chemical parameters that will be required to calculate the accumulation and concentration in roots and aerial vegetation are the partition coefficients, between octanol (O), air (A), and water (W), namely K_{OW} and K_{AW} or K_{OA}, and photodegradation and biometabolism reaction rates. For more information on modeling uptake of chemicals by vegetation, refer to Trapp *et al.* (1994a, 1994b) and Trapp and McFarlane (1995).

4.10 Soil Compartment Treatment

Because soil is not a well-mixed compartment such as air or surface water, the model developers recognized that the simple boundary-layer approach for describing mass transfer from soil to air will not always be applicable because the concentration of chemicals in soil may not be considered "well-mixed" till a depth of several meters. By contrast, in the air and water compartments, the boundary layer thickness that defines "well-mixed" is in the order of millimeters. Thus, there is a need to develop better ways of modeling chemical concentration in soils and to determine the appropriate depth of the surface layer of soil.

There have been several approaches to the problem of devising a simple but accurate model of soil transport. The approach used in three of the models (ChemCAN, HAZCHEM, and SimpleBOX) is to use a fixed diffusion-path length of typically 0.05 m, independent of the chemical being evaluated. Because these models have a relatively thin soil layer (0.1 m), they cannot be used to predict the fate of chemicals that have been incorporated into soil at depths of several meters, as is the case at hazardous waste sites. Mackay (1991) observed that using a single soil layer and a half-depth boundary layer may significantly underestimate volatilization at the soil surface. He suggested two potential remedies for this situation:

- use more than one soil-layer compartment in the multi-media model, and
- use the geometric-mean value of the soil-layer depth as an approximation for the boundary layer depth.

One approach to determining an appropriate soil-layer depth is to use the *limiting soil-cover depth*, which is the thickness of soil required to limit volatilization loss over an infinite time to less than 1% of the initial concentration. At that depth, volatilization and/or deposition at the surface will have essentially no impact on concentration. At shallower depths, there will be a gradient of concentration that will be controlled by the rate of chemical deposition and/or evaporation at the surface of the soil. Jury *et al.* (1990) determined that the limiting soil-cover depth is dependent on the characteristics of the chemical and varies in sandy soil from 0.001 m for chrysene to 160 m for dichlorodifluoromethane.

The CalTOX model uses a different approach to treat chemical concentration in soil. Soil layers are designated to represent the zone between the soil surface and the top of the saturated zone, and a regression model is developed that uses effective soil diffusion coefficients to estimate diffusion depth in each of these soil layers. This regression equation was developed by conducting multiple simulations of this soil system using the model of Jury *et al.* (1983) and a range of chemical properties (K_d and K_{AW}) and effective soil diffusivities. The regression equation was then determined by an optimized fit against the results of these simulations. In effect, this procedure allows CalTOX to mimic the Jury *et al.* (1983) model with a simple regression model based on effective diffusivity. This is similar to the approach recommended by Cohen *et al.* (1990), that a spatial diffusion model be used to represent the soil compartment in a multi-media model.

An alternative suggested by the model developers for harmonizing these two different approaches is to introduce a chemical-dependent "effective soil depth" equal to the penetration depth of the chemical. The classical Damköhler number, D_a, (Damköhler, 1937) can be used as the theoretical concept to determine a reasonable penetration depth for chemicals into surface soils, where this penetration depth is the depth at which, at steady-state, the rate of chemical reaction-disappearance is equal to its rate of movement into soil by diffusive processes. When these rates are equal, D_a is unity, 1, and the penetration depth, d, is calculated from $(D/k)^{\frac{1}{2}}$ where k is the first-order reaction rate and D is the molecular diffusivity.

The model developers recommended that instead of this classical form of the Damköhler number, a modified Damköhler number that incorporates the adsorptive penetration by diffusion plus advection into the upper soil layer be used to calculate the penetration depth. In this related formulation, the ratio of the combined reaction rate in both the pore water and soil solids and the transport depth are used to calculate the modified Damköhler number, D_a^*

$$D_a^* = rRd/[\mu/\varepsilon + D/d]C_s$$

where r is combined rate of reaction, R is the retardation factor, μ is the Darcian velocity into the soil, ε is soil porosity, and C_s is the soil pore water concentration. When r is assumed to be equal to kC_s and $D_a = 1$, the penetration depth, d, is calculated by

$$d = [\mu/\varepsilon + \sqrt{(\mu/\varepsilon)^2 + 4DkR}\,]/2kR$$

Note that this algorithm cannot be used to estimate the soil penetration depth for non-degradable chemicals because for these chemicals, k = 0 and d is undefined. In such cases, other reasonable and practical considerations will be required to estimate d.

The approach described above should be used to derive a chemical-dependent soil-layer depth except when mechanical mixing of the soil layer has occurred, as in agricultural soils. In this case, the soil depth should be taken as the actual mixing depth, or 20 cm as a default value. If the computed penetration depth is less than realistic estimates of the actual weathered soil thickness, then the computer penetration depth should be used. If not, then the actual weathered soil thickness or a default value of 20 cm should be used.

There is a clear need to test these and possibly other approaches with a view to establishing a preferred method of treating the fundamental heterogeneity of soils.

4.11 Aquatic Sediments Treatment

Of all the environmental compartments included in multi-media models, the aquatic bottom sediment is possibly the least well known and includes numerous mechanisms of chemical transport, most of which are poorly understood (Reible *et al.*, 1991). Our ability to specify appropriate transport coefficients between sediments and overlying water for the various water bodies representing the UGU in Table 4-2 is inadequate.

To help in the representation of sediments, the model developers recommended classifying sediments into two categories based on active water flow and particle transport. In the first category, transport is rapid and the capacity for chemical accumulation is small. Typically, these sediments occur in fast flowing streams and consist primarily of sand and small boulders. The stream water actively circulates through the top 10 cm of the bed. The particles of large size present less surface area for chemical sorption. Many seasonal streams with high spring-time flows fall into this category. In this category of sediments, any organisms and chemicals that accumulate during low flow period are typically transported out of the system during storm events or spring floods.

The second category represents aquatic systems of moderate to low flow where particle accumulation occurs. In these systems, the sediment compartment will have a moderate to large capacity for chemical. The transport coefficient between the sediment and overlying water can be quite variable and will depend upon the presence and type of organisms in that sediment. Bioturbation and molecular diffusion represent the extremes of the range of transport coefficients. Large populations of organisms which move active sediment and pore water in the top 5 to 10 cm of the sediment will result in large transport coefficients.

The model developers recognize that the subject of chemodynamics in bottom sediments is a very active research area, and as new information becomes available, the way sediment processes are represented in these multi-media models will improve. For example, it is well known that a nepheloid layer exists on the water side. Research is focusing on how this layer, consisting of flocculated particles high in organic matter, should be represented in models. Other areas of research include how organisms migrating in and out of the sediments affect chemical fate, how metabolism of organic chemicals by these organisms should be represented, and the extent to which the generation of dissolved organic carbon (DOC) within the bed and its subsequent release to the water column is a major transport mechanism for hydrophobic chemicals.

In the opinion of the model developers, it is likely that in most of the UGU, only one or perhaps two of these transport processes from the sediment will strongly influence the exchange of chemicals across the sediment-water interface. It is hoped that ongoing research will determine which processes are significant and under what conditions, so that within a few years the uncertainty in representing and parameterizing the transfer processes between sediments and overlying water will be reduced significantly.

5: Accuracy And Validation of Multi-Media Models

When mathematical models are considered for use in regulatory applications, especially when quantitative results are required, these key questions always arise: "Is the model validated?" and "How accurate are the model results?" The issue of validation and model accuracy is complex and contentious (Oreskes *et al.*, 1994); however, there is a need continually to subject models to both verification and validation exercises. A verification exercise is intended to confirm that the model does what the modeler intends, *i.e.*, it is mathematically or computationally correct. A validation exercise is intended to confirm that the model accurately represents the process it is designed to mimic. The primary intent of the exercise conducted as part of this workshop was to verify that a set of multi-media models, which were all constructed from the same fundamental premises of mass balance, yielded similar results when given the same input parameters. This exercise has been useful as a determination of precision among a set of current multi-media models. The issue of accuracy or validation involves comparison of model behavior relative to actual measured values of chemical fate in a regional environment. This type of exercise has as yet not been carried out with a high degree of sophistication.

5.1 Perceived Multi-Media Model Accuracy

The accuracy of multi-media models is determined by the accuracy of the model inputs; the accuracy with which the model addresses the relevant physical, chemical, and biological processes; and the model resolution (both in space and time). Therefore, absolute values for accuracy cannot be easily generated. Nevertheless, to guide prospective users on the level of accuracy that could in general be expected from multi-media models, the model developer group compiled Table 5-1. This table represents their consensus on the accuracy of chemical concentration estimates. Although these factors are subjective and based on professional judgment rather than on hard statistical quantities, the model developer group believes that these factors correctly reflect the presently perceived accuracy of state-of-the-art multi-media models.

TABLE 5-1 *Perceived Accuracy of Multi-Media Model Predictions of Concentration*

Chemical Type	Source Medium		Primary Sink Medium	Secondary Sink Medium
	Air or Water	Soil		
Remains in media to which discharged	2	3	not relevant	not relevant
Properties well known	3	5	further 2x	further 2x
Properties uncertain	20	30	further 5x	further 5x

The perceived accuracy factors in Table 5-1 are based on the assumptions that the emission rate is known exactly, that the degradation rates are well characterized, and that only the uncertainties in chemical properties and chemical transfer between media determine the actual predicted concentrations. The accuracy factors in the table are further discriminated based on three chemical types and three categories of media. The chemical types are those

- which tend to remain primarily in the medium of discharge because of extreme values in physical-chemical properties (vapor pressure, solubility, or sorption, *e.g.*);
- which have well-established properties, but may be subject to multi-media partitioning; or
- for which key properties are uncertain, or existing algorithms for expressing transport or transformation processes may not be reliable.

The media are categorized into

- source medium, *i.e.*, the medium into which discharge occurs;
- primary sink medium into which the chemical migrates from the source medium; and
- secondary sink medium into which there is subsequent migration from the primary sink medium.

For example, if PAH is discharged to air, then air is the source medium, water is the primary sink medium, and sediment is the secondary sink medium. The reason for this distinction is that for each inter-media transfer the uncertainty in predicted exposure concentration increases. A factor of 3 for a given medium implies that an estimated concentration of 12 will have 95% confidence limits of 4 and 36, such that the true average environmental concentration will most likely lie between these limits. It is recognized that environmental concentrations vary, thus this table should be viewed as providing only an approximation of the accuracy. These factors assume that the discharge of the chemical into the source media is known exactly. In practice, the uncertainty in the multi-media model results may be much larger due to considerable uncertainty in the emission rates.

5.2 Evaluations of Multi-Media Model Accuracy

Although there are very few validation exercises that can confirm the values in Table 5-1, there has been progress in this regard. Technical Report 58 on the Non-Occupational Exposure to Chemicals (ECETOC, 1994b) suggests that for di-2-ethylhexylphthalate (DEHP) and benzene, the HAZCHEM model gives results which are in accord with observations as shown in Table 5-2. There is always a possibility of fortuitous agreement and cancellation of errors, thus there is a need to extend these validations to more chemicals and more locations. This may apply to DEHP because of uncertainties in its octanol-water partition coefficient.

TABLE 5-2 *Comparison of Measured Environmental Concentrations with Calculated Concentrations of DEHP for Different Environmental Compartments*

Medium	Measured	Reference	Calculated (HAZCHEM)
Air (mg/m^3)	$0.4 - 2.9 \times 10^{-6}$ $< 5 \times 10^{-6}$ (remote areas)	1 2,3	0.6×10^{-6}
Water (mg/L)	$0.05 - 30 \times 10^{-3}$ $< 0.1 - 4.0 \times 10^{-3}$ (rivers, lakes) up to 3.0 (rivers, lakes) $0.05 - 1.4 \times 10^{-3}$ (groundwater at waste water infiltration sites)	1 2 3 3	1.2×10^{-3} (surface water) 0.13×10^{-6} (groundwater)
Sediment (mg/kg)	1 - 70 (river sediment) (up to 1500 sediment near discharge points) 1.8 - 18.3 (Rhine) 0.1 - 8.9 (Weser)	2 2 4 4	0.7
Soil (mg/kg)			0.1×10^{-3} (arable) 0.22×10^{-3} (natural)
Food (mg/kg)	0.01 - 10 (fish) < 1 (US food survey, 1974)	2 3	0.8×10^{-3} (vegetables) 0.2×10^{-3} (meat) 1.4×10^{-3} (milk fat) 4.4 (fish) 3.5×10^{-5} (drinking water)
Daily intake (mg/kg)	$4.5 - 30 \times 10^{-3}$	2	2.3×10^{-3} (adult) 2.8×10^{-3} (child)

Sources: 1. International Agency for Research on Cancer (IARC), 1982
2. International Programme on Chemical Safety (IPCS) 1992b
3. UK Ministry of Agriculture, Fisheries and Food (UK-MAFF) 1987
4. Furtmann, 1993

5.3 The Need for Model Validation

To improve the validation and to determine the accuracy of multi-media models, additional validation studies are needed. However, it must be emphasized that validation can relate only to a specific result for a specific set of conditions. For example, a statement of validation could read "If the emission rate into environmental medium, N, of a given region with a specified set of characteristics is equal to E, then the concentration in environmental medium, M, will be equal to C." The validity of the predictions for each set of conditions must be addressed separately. Furthermore, the definition and extent of validation may be different for each type of model application. Thus, "validity of the model" cannot be ascertained in an absolute or universal sense. However, as more experience is gained with successful applications, the model will become more universally accepted as yielding valid results.

The objectives of a validation exercise include

- assessing the expected precision and accuracy in different situations for discharge of different classes of chemicals, and
- indicating in which situations and for which classes of chemicals the models do not give sufficient and acceptable precision and accuracy for the area of application.

Based on these objectives, an approach for validating multi-media models would including the following steps:

- Select five to ten "representative" chemicals with "good" field data sets of measured environmental concentrations.
- Calculate environmental concentrations and attendant uncertainty using several models.
- Compare the distribution of calculated and measured concentrations for each medium.
- Determine the model's accuracy and suggest possible improvements.

This last step illustrates model development's iterative nature. As discrepancies between calculated and measured concentrations become apparent, this information should be used to further refine the model.

The data required to conduct such a validation study include

- production and release data (*e.g.*, ton/y). Data should be for a specified area (*e.g.*, Germany, Europe). Any advective inflows must also be specified.
- chemical characteristics data. This includes physical-chemical data and reaction rates (*i.e.*, half-lives in different compartments).
- environmental characteristics and monitored concentrations. Preferably, this information should be from several compartments such as air, water, sediment, soil, food (*e.g.*, fish), vegetation, with emphasis on the primary compartment of accumulation.

In choosing chemicals for validation, it is recommended that a validation exercise include at least one chemical from each of the following five classes:

- volatile or partially volatile organic chemicals (*e.g.*, chlorobenzenes, PCBs, PAHs)
- involatile chemicals (*e.g.*, metals, surfactants)
- insoluble chemicals (*e.g.*, chloroparaffins, silicones, waxes)
- dyes and polymers (*e.g.*, involatile and insoluble)
- fast-reacting chemicals (*e.g.*, mercury, phenols, hydroquinone)

Validation data sets must be carefully selected, giving due consideration to their accuracy and general applicability. Unfortunately, accurate emissions data are rarely available; therefore, data sets of the highest possible emission accuracy should be sought and selected.

6: Improving the Models: Research Needs

The following were identified as the most critical research needs which, when satisfied, should result in an improvement in the performance of models. The list is not intended to be exhaustive.

6.1 Chemical Properties, Classification and Emission Data

There is need to measure and correlate chemical properties, especially over a range of temperatures and in realistic environmental conditions, for example, in the presence of electrolytes. More reliable methods of averaging chemical properties over, for example, a range of temperatures should be developed.

More reliable estimates should be obtained of transformation half-lives and their dependence on environmental conditions. There must also be better guidance on how laboratory studies should be used to estimate environmental half-lives. Test methodologies should be developed to determine primary biodegradation rate constants.

More and better emission data and background environmental concentrations are required, since these quantities are the most critical input when models are used to deduce concentrations.

6.2 Environmental Characteristics

There is a need to develop default data sets for a variety of UGU. The incorporation of multi-media models into GIS should also be encouraged and demonstrated.

There has been a fair amount of laboratory research and field simulation work devoted to measuring and correlating the inter-media transfer coefficients. These data should be compiled, evaluated, and condensed for use in describing the various environmental types or units. A research/study project should be undertaken to recommend average numerical values of transport coefficients, including temperature dependence, with ranges, for incorporation into models.

6.3 Sensitivity and Uncertainty Analysis Methods

There should be a coordinated program to develop systematic uncertainty and sensitivity analysis methodology that can be incorporated into modeling frameworks.

There is also a need for guidance on how to interpret the results of uncertainty and sensitivity analyses.

6.4 Improved Model Formulation

A number of areas were identified in which there is a need to improve model performance and reliability.

Processes of uptake by vegetation should be studied with a view to developing simple, robust, but reliable models.

The effects of snow and ice on the environmental fate of chemicals require study.

Improved models for chemical fate in soil and sediments are required. The incorporation of waste water treatment models into emission estimation is encouraged.

The process of bioaccumulation in animals and plants proves to be a critically important component of the assessment of chemical fate. There is thus a need to improve the interface between multi-media models and bio-uptake models which can properly take into account processes such as uptake of chemicals by vegetation, transport in the agricultural and wildlife food chains, the effects of aquatic food chain structure, metabolic conversion, and the determinants of absorption efficiency for a variety of chemicals.

6.5 Model Comparison

To ensure precision among models used by different agencies, performance criteria should be developed. Models should be periodically evaluated against these criteria in international round robin exercises. There should be provision for periodic review and updating of these criteria as the science improves. To ensure broad application of these performance criteria, they should be developed by a group that includes both users and developers.

In addition, the participants recommended that a set of criteria be developed and implemented for documenting any model that is to be used in a regulatory application (see Appendix D).

6.6 Model Validation

Data should be located and compiled for model validation exercises. Possible data sets identified by the participants include data on naturally produced chemicals (*e.g.*, bromoform, which is produced in large quantities by marine algae) and measured concentration changes in several environmental compartments following changes in discharge rates as a result of bans or regulation (*e.g.*, lead, PCBs, and certain pesticides). In this respect, studies of sediment chronology will be useful in reconstructing historical emission patterns. Applicable data sets can be obtained from reports such as "Daten zur Umwelt" edited by Umwelt Bundesamt, "Handbuch Umwelt Chemikalien" by G. Rippen, and many others (*e.g.*, RIVM reports, Danish Environmental Protection Agency, Japan).

Comparison should be made between measured and predicted concentrations for several data sets. Ideally, this should be done by groups of experts acting independently and using different models.

Appendix A: Workshop Participants

Name	Address	Phone Numbers	Affil.	Modeling Expertise
Dr. Robert S. Boethling	Office of Polution Prevention and Toxics (7406) U.S. Environmental Protection Agency 401 M. Street S.W. Washington, D.C. 20460, USA	T: 202-260-3912 F: 202-260-0981 e-mail: boethling.robert @ epamail.epa.gov	Gov.	User
Dr. David N. Brooke	Building Research Establishment Bucknall Lane, Garston Watford, WD2 7JR U.K.	T: 44-192-366-4862 F: 44-192-366-4609	Gov.	Expert User
Dr. Davide Calamari	Istituto di Entomologia Agraria Universitá di Milano Via Celoria 2 I-20133 Milano Italy	T: 39-2-236-2880 or 3439 F: 39-2-266-80320	Acad.	Model Developer
Dr. Christina E. Cowan	The Procter & Gamble Company Ivorydale Technical Center 5299 Spring Grove Ave. Cincinnati, OH 45217-1087, USA	T: 513-627-8216 F: 513-627-8198, 1354 e-mail: cowance@pg.com	Ind.	Model Developer and Expert User
Dr. Jack H.M de Bruijn	Ministry of Housing, Spatial Planning and the Environment DGM/SUS, IPC 655 P.O. Box 30945 2500 GX The Hague The Netherlands	T: 31-70-339-4845 F: 31-70-339-1314 e-mail: debruijn @ dsvs.dgm.minvrom.nl	Gov.	Expert User
Dr. James Devillers	CTIS 21 rue de la Banniere 69003 Lyon France	T: 33-7-862-8499 F: 33-7-862-9912	Priv. Res. Lab.	Model Developer
Dr. Tom Feijtel	Procter & Gamble ETC Temselaan 100 B-1853 Strombeek-Bever Belgium	T: 32-2-456-2329 F: 32-2-456-2845	Ind.	Model Developer and Expert User
Dr. Nicola Grandy	OECD, Environmental Health & Safety 2 Rue Andre-Pascal 75775 Paris Cedex France	T: 33-1-452-41676 F: 33-1-452-41675	Gov.	User

Dr. Bjorn Hansen	Commission of the European Communities Joint Research Centre Environmental Institute European Chemicals Bureau I-21020 Ispra VA Italy	T: 39-332-78-5884 F: 39-332-78-9963 e-mail: b_hansen@jrc.it	Gov.	Model Developer
Dr. Toshiyuki Hattori	Technology Department Integrated Research Division Japan Research Institute, Limited 16 Ichibancho Chiyoda-ku Tokyo, 102 Japan	T: 81-3-3288-4142 F: 81-3-3288-4689	Priv. Res. Lab.	User
Prof. Dr. Sven E. Jørgensen	DFH Environmental Chemistry University Park 2 2100 Copenhagen Denmark	T: 45-353-70850 F: 45-353-75744	Acad.	Expert User and Model Developer
Dr. David Kane	Health and Welfare Canada Environmental Health Centre Tunney's Pasture Ottawa Ont. K1A 0L2 Canada	T: 613-957-9279 F: 613-954-2486 e-mail: dkane@upb.hwc.c a	Gov.	User
Dr. Donald Mackay Dr. Antonio Di Guardo Dr. Frank Wania	Institute for Environmental Studies University of Toronto Toronto, Ontario M5S 1A4 Canada	T: 416-978-4019 F: 416-978-8605 e-mail: cec@ecf.utoronto. ca	Acad.	Model Developer
Dr. Neil Mackay	Corning Hazleton Otley Road, Harrogate North Yorkshire HG3 1PY U.K.	T: 44-1423-500011 F: 44-1423-525620 e-mail: nmackay@hazle.c o.uk	Ind.	Model Developer
Prof. Dr. Michael Matthies	University of Osnabrück Institute of Environmental Systems Research D-49069 Osnabrück Germany	T: 49-541-9692576 F: 49-541- 9692599 e-mail: Matthies@ramses.u sf.Uni-Osnabrueck.de	Acad.	Model Developer
Dr. Thomas McKone Ken Kiefer Randy Maddalena	Risk Sciences Program Environmental Toxicology Dept. University of California Davis, CA 95616. USA	T: 916-754-8757 F: 916-752-3394 e-mail: mckone1@llnl.gov	Acad. and Gov.	Model Developer
Dr. Thomas F. Parkerton	Exxon Biomedical Sciences, Inc. Mettlers Road, CN-2350 East Millstone, N.J. 08875-2350, USA	T: 908-873-6367 F: 908-873-6009	Ind.	Model Developer and Expert User

Dr. Richard Purdy	3M Company Environmental Laboratory P.O. Box 33331 St. Paul, MN 55133, USA	T: 612-778-5379 F: 612-778-6176 e-mail: repurdy@mmm.co m	Ind.	Model Developer and Expert User
Dr. Richard Stephenson	Shell International Petroleum Maataschappij B.V. P.O. Box 162 2501 AN The Hague The Netherlands	T. 31-70-377-6125 F: 31-70-377-6204	Ind.	User
Dr. Roger Sutcliffe	Commercial Chemicals Evaluation Branch Environment Canada 351 St. Joseph Blvd. Hull, Quebec K1A 0H3, Canada	T: 819-953-2461 F: 819-953-7155	Gov.	User
Dr. Wil F.J.P. ten Berge	Corporate Safety, Environment, Health & Technology P.O.Box 6500 6401 JH Heerlen The Netherlands	T: 31-45-787128 F: 31-45-787112	Ind.	Model Developer and Expert User
Dr. Louis J. Thibodeaux	Dept. of Chemical Engineering Louisiana State University Baton Rouge, LA 70803-6421 USA	T: 504-388-1426 F: 504-388-1476 or 5990	Acad.	Model Developer
Dr. Dik van de Meent	Laboratory for Ecotoxicology National Institute of Public Health and Environmental Protection Antonie van Leeuwenhoeklaan 9 P.O. Box 1 720 BA Bilthoven The Netherlands	Tel: 31 30-74911 Fax: 31-30-251925 e-mail: ecodm@rivm.nl	Gov.	Model Developer
Dr. M.M.H.E. van den Berg	Health Council of the Netherlands Prinses Margrietplantsoen 20 The Hague The Netherlands	T: 31-70-3441860 F: 31-70-3837109	Gov.	User
Dr. Dolf van Wijk	Akzo Nobel Research Lab. of Ecotoxicology P.O. Box 9300 6800 SB Arnhem The Netherlands	T: 31-85-665401 F: 31-85-662528	Ind.	Model Developer and Expert User
Dr. William Veerkamp	Shell International Petroleum Maataschappij B.V. P.O. Box 162 2501 AN The Hague The Netherlands	T: 31-70-377-1820 F: 31-70-377-6204	Ind.	Model Developer and Expert User

Note: Affil. = affiliation; Acad. = academia; Ind. = industry; Gov. = government; Priv.Res.Lab. = private research laboratory

Appendix B: Terminology

Throughout this report, certain terminology is used when describing model features. These concepts may not be familiar to all readers, thus we review briefly some of these key concepts.

By **media,** we imply phases of the environment such as air, water, soil, sediments, and biota which are generally similar in internal properties. A medium may contain other media within it, for example, fish and suspended matter in water. The words *phases* and *compartments* are used almost synonymously.

Equilibrium implies thermodynamic equilibrium between media, *i.e.*, a common chemical potential or a state which is reached after prolonged contact. The ratio of these concentrations is a **partition coefficient**. For example, an air-water partition coefficient for benzene is about 0.3, implying that a concentration of 0.3 in air and 1 in water are in equilibrium.

Steady-state implies unchanging with time. A polluted river may contain dissolved oxygen at a constant 2 mg/L and is thus at steady-state. It is not, however, in equilibrium with the atmosphere since this concentration would be about 8 mg/L. **Unsteady-state** implies changing with time or dynamic.

Media are often referred to as **well mixed** or **homogeneous**. This implies that concentrations are assumed to be identical at all points, thus only one concentration need be defined. This greatly simplifies the mathematics of mass balancing. Shallow lakes and the lower atmosphere are sometimes well mixed because of the prevalence of mixing currents.

Transport processes convey chemical from one medium to another; for example, evaporation is a **diffusive process** which occurs when disequilibrium exists. **Advective processes** convey chemical in currents of air and water, or in association with moving particles.

Transformation processes result in a substance changing its chemical structure, *e.g.*, benzene to phenol by a chemical reaction. **Mineralization** represents the entire process of conversion of an organic compound to CO_2 and water. **Degradation** is often synonymous with transformation.

Some chemicals exist as several **species**, for example, acetic acid and acetate ion. There is usually a potential for reversible conversion, as distinct from transformation which is usually irreversible.

A **mass balance** is an axiomatic assertion that a quantity of chemical must be fully accounted for, *i.e.*, it does not form or vanish spontaneously. It forms the basis of the mass balance equation which is applied to a medium in the form

Inventory change = Input rate - Output rate (such as units g/hour).

In a **closed system** there is no input or output of chemical, whereas in an **open system** there is input or output of chemical.

When assembling mass balance models, various numbers of media can be included: 1 or 2 or 3 soil layers, for example. Steady- or unsteady-state conditions may apply. It has become common practice to express certain of these conditions as **levels** of the model.

A **Level I model** is a closed system mass balance of a defined quantity of chemical as it partitions at equilibrium between compartments. There is no reaction. Steady-state and equilibrium apply.

A **Level II model** is a steady-state open system description of chemical fate at equilibrium with constant chemical emission rate. It includes reaction and advection as loss processes. It is also at steady-state and equilibrium.

A **Level III model** is a steady-state description of chemical fate between a number of well-mixed compartments which are not at equilibrium. This level thus includes inter-media transport expressions.

A **Level IV model** is an unsteady-state version of Level III. Level II can also be written in unsteady-state form.

For further explanation of these terms and illustrations of these concepts, refer to texts such as those of Thibodeaux (1979), Mackay (1991), or Jørgensen (1994).

Appendix C: Questionnaire Responses

Application of Multi-Media Fate Models to Regulatory Decision Making

1. **What experience do you or your agency have with application of multi-media models?**

   Japan: Introducing "CalTOX" Japanese version to Japan

   U.S.: The office of pollution prevention and toxics has developed two multi-media models, ENPART (1982) and TOXSCREEN (1984). Used in exposure assessments in early 1980s. Steady-state modeling of PCBs also in early 1980s.

   Netherlands: Considerable experience in application for priority setting and risk assessment.

   Canada H&W: Development of Canadian-specific, fugacity-based modes by D. Mackay over the last 5 years.

   Denmark: Multi-media models are being developed that will be used by agencies for chemical assessments mostly on pesticides.

   UK: Have used multi-media (Mackay) model on individual chemical assessments in past, beginning to use them more formally within the EU and OECD programmes. Model use has included developing compartments dimensions and properties relevant for UK.

2. **If you or your agency have not used multi-media models, why?**

   Japan: Have water and soil quality criteria which can be used to set clean-up goals.

   U.S.: Multi-media models have seen limited use under TSCA for these reasons: (1) emphasis is on media into which chemical is discharged, therefore, focus is on near-field concentrations; (2) long range issues have typically been dealt with by other EPA programs.

3. **If you or your agency have used multi-media models, indicate the frequency with which you use multi-media models for each of the types of applications on the list below. If possible give specific examples in the space provided.**

   - Estimating exposure concentrations in environmental media for risk assessment

25% or less	50%	75%	100%
3		1	1

   - Ranking and or prioritizing of chemicals

25% or less	50%	75%	100%
3	1	1	

- Estimating exposure to humans and/or higher animals via food chain

25% or less	50%	75%	100%
2		1	1

- Defining hazardous waste site clean-up goals

25% or less	50%	75%	100%
1		1	

2, no response or unsure

- Contribute to or verify general understanding of a chemical's fate

25% or less	50%	75%	100%
1	1	1	2

- Environmental quality criteria (*e.g.*, water quality or soil quality) setting

25% or less	50%	75%	100%
1	1		

3, no use

- Spill assessments

25% or less	50%	75%	100%
2			

3, no use

- Other — Specify

  No other application areas identified.

4. **Which of the following potential application areas do you see as the emerging or future regulatory issues that multi-media models could be applied to and the level of importance of having multi-media models to address this need?**

- Estimating exposure concentrations in environmental media for risk assessment

Never	Small Potential	High Potential	Definitely
		3	3

- Ranking and or prioritizing of chemicals

Never	Small Potential	High Potential	Definitely
	2	1	3

- Estimating exposure to humans and/or higher animals via food chain

Never	Small Potential	High Potential	Definitely
	2	2	1

- Defining hazardous waste site clean-up goals

Never	Small Potential	High Potential	Definitely
1	3	1	

- Contribute to or verify general understanding of a chemical's fate

Never	Small Potential	High Potential	Definitely
		4	1

- Environmental quality criteria (*e.g.*, water quality or soil quality) setting

Never	Small Potential	High Potential	Definitely
2	3	1	

- Spill assessments

Never	Small Potential	High Potential	Definitely
2	2	1	

1, no opinion

5. What types of chemicals do you currently apply multi-media models to? (Indicate top two types.)

Chemical Type	Number of responses
nonpolar non-ionizable organic	6
surface-active organic	1
ionizable organic	1
metals	3
water soluble polymers	0
virtually insoluble hydrocarbons or polymers	0

6. What types of chemicals would you like to or anticipate in the future needing to apply multi-media models to? (Indicate top two types.)

Chemical Type	Number of responses
nonpolar non-ionizable organic	3
surface-active organic	1
ionizable organic	1
metals	4
water soluble polymers	2
virtually insoluble hydrocarbons or polymers	1
all of the above	1

7. **What is the expected expertise and background of the model users?**
   - Degree B.Sc. MST. Ph.D.:
   - years of operation of models (*i.e.*, multi-media or other):
   - years of experience in environmental chemistry:
   - years of experience in risk assessment:

   Users will range from B.Sc. to Ph.D. with wide range of experience in application of models, environmental chemistry, and/or risk assessment.

   One respondent stated that environmental chemistry was the most important experience base.

8. **How much training do you expect the users to be able to get and would this have to be in-house or could it be outside workshops that will require travel? What are constraints to training, *i.e.*, time, money or other (please explain).**

   Most training would have to be in-house and probably on-the-job because of constraints in travel money.

9. **To what extent do you feel that multi-media models used in regulatory decision making should be harmonized?**

   Five respondents said harmonization was important, one said no, and there was one no response.

   Reasons given in support of harmonization:
   - quality control
   - in earliest stages of risk assessment to allow communication of results among countries that are working together
   - ranking of chemicals can be done on a common basis

   Reasons given against harmonization:
   - not appropriate for regional and site-specific assessments
   - better to share results of common examples than to rely on harmonized models

   environmental compartments included yes no
   3, yes

   endpoints considered yes no
   1, yes

   algorithms that are used to describe the environmental transfers yes no
   2, yes

   compartment sizes and characteristics yes no
   3, yes

10. If you use or anticipate using multi-media models for estimating exposure concentrations in environmental media for risk assessment, what are your

- desired outputs (*e.g.*, chemical distribution, single concentration in each environmental compartment, range of potential concentration)

  All responses, yes

- acceptable uncertainty in each of the endpoints (*e.g.*, within factor of 2, within order of magnitude)

  Would like uncertainty factor of 5 or less, but could live with 10 or less.

- level of detail and number of compartments/media used (*e.g.*, several soil compartments to represent different types or depth of soil)

  Minimum model will include soil, air, water and sediment. Actual number of compartments will depend on the application area. Important that number of compartments (*i.e.*, complexity) and level of uncertainty be balanced.

- number or transport, mass action processes considered (*e.g.*, aerosol deposition, sorption)

  Dependent on number of compartments, but should include all for the compartments included in the model

11. If you use or anticipate using multi-media models for ranking and or prioritizing of chemicals,

Answers were the same as in 10 with addition of comparison to effect concentration data.

12. If you use or anticipate using multi-media models for estimating exposure to humans and/or higher animals via food chain,

Answers were the same as in 10 with addition of concentrations in food and higher organisms. Therefore model will need to include vegetation, fish and higher organisms.

13. If you are using or anticipate using a multi-media model for defining hazardous waste site cleanup goals,

Answers were the same as in 10 with addition of requirement for potential ranges in concentrations. Also may require addition of compartments for food and higher organisms depending on endpoint used to establish goals.

14. If you are using or anticipate using a multi-media model to contribute to or verify general understanding of a chemical's fate.

Answers were the same as in 10.

15. **If you are using or anticipate using a multi-media model for setting environmental quality criteria (*e.g.*, water quality or soil quality),**

    Answers were the same as in 10 with addition of comparison to effect concentrations data.

16. **If you are using or anticipate using a multi-media model for spill assessments,**

    Answers were the same as in 10 and 12, depending on the desired assessment endpoint.

17. **If you are using or anticipate using multi-media models for other applications,**

    No other applications identified

18. **What other issues or concerns do you have around using multi-media models in regulatory decision making?**

    Who can authorize/certify multi-media models and what are the criteria? How can these authorized models get public acceptance?

    Errors in properties and degradation rates propagate, such that order of magnitude certainty can be difficult to achieve. This is of critical concern. Often uncertainty with respect to parameters such as biodegradation rate is so high that the value of multi-media modeling can legitimately be questioned.

    Time Scales: The steady-state assumption is not good enough for many applications

    Uncertainty analysis: what variation can be expected and to what parameters/properties can the uncertainty be ascribed?

    Dependence of output of models on regional variation (*i.e.*, temperature, compartment size, *etc.*)

    Multi-media models are excellent tools for selection of chemicals that are more acceptable from an environmental point of view. It is excellent for ranking and understanding of where the weak points are in the chain. They cannot be used for predictions and a specific site, but only give rough regional estimates. They should never be over-sold.

Appendix D: Criteria for Model Documentation

Documentation

Documentation for all programs should

- be referenced with each model's name, version, date, and organization of development.
- be user friendly.
- contain a full description of each basic model, including its purpose and scope.
- include source code and/or a full description of all algorithms used in each model.
- include instructions on version control, describing modifications made to earlier versions of each model.
- contain a description of the inherent assumptions, limitations, and individual model components.
- describe the parameters' input and structure.
- fully reference the appropriate original reference.
- include hardware and software specifications, and describe all steps required for installation.
- contain information on the operation of each model, including input/output structures and calculation procedures.
- describe any sensitivity and uncertainty analyses.
- contain a full description of available validation studies.
- include in the user's manual a discussion on the spatial dimensions of each model contained in the program.

Model Operation (User Friendliness)

The programs should

- be easy to operate.
- use uniform input format (International System Of Units [SI]).
- allow easy importation of data sets (variable components).
- afford transparency of the generic data sets (fixed components).
- have an output format that includes absolute quantities, concentrations, and percentages of material in each compartment, where appropriate.
- have an output format available in graphic and/or numerical formats.

QA/QC

- All output should be properly documented to allow verification against input data sets.
- Output documentation should provide the following information:

1. Identification of the program, version, date
2. Input parameters used (user ID, date, substances data, scenarios)
3. Output data

- A description of the hardware and software required to rebuild the executable program, including the development environment (languages, systems, and other required utilities) and source code, should be archived at the development organization.
- Verification data sets (input and output) should be provided.
- Advice should be provided for input in the case of data gaps.
- Input parameters should be "range-protected" wherever possible.
- Documentation of the peer review carried out on each model should be provided and should include a source code check.

Support

Support for the program should include

- a contact and distribution organization.
- information, help, and support facilities including mail and/or e-mail address, phone and fax numbers.
- clear designation of program version.
- the facility for training of users where possible and appropriate.
- a user registration process.

References

Bacci, E., D. Calamari, C. Gaggi and **M. Vighi.** 1990. Bioconcentration of organic chemical vapors in leaves: Experimental measurements and correlations. *Environ. Sci. Technol.* **24**:885-889.

Baker, J.E., S.J. Eisenreich and **D.L. Swackhamer.** 1991. Field measured associations between polychlorinated biphenyls and suspended solids in natural waters: An evaluation of the equilibrium partitioning paradigm. In R.A. Baker, (ed.), *Organic Substances and Sediments in Water.* Lewis Publishers, MI, pp. 79-89.

Boesten, J.J.T.J. 1991. Sensitivity analysis of a mathematical model for pesticide leaching to groundwater. *Pestic. Sci.* **31**:375-388.

Briggs, G.G., R.H. Bromilow and **A.A. Evans.** 1982. Relationships between lipophilicity and root uptake and translocation of non-ionized chemicals by barley. *Pestic. Sci.* **13**:495-504.

Brooke, D. and **P. Matthiessen.** 1991. Development and validation of a modified fugacity model of pesticide leaching from farmland. *Pestic. Sci.* **31**:349-361.

Bureau, E.C. 1994. *Priority setting using the IPS method: From IUCLID to IPS ranking.* Doc. XI/ECB/8883/93 (Rev. 1). Joint Research Centre, Ispra, Italy.

Clark, K.E., F.A.P.C. Gobas and **D. Mackay.** 1990. Model of organic chemical uptake and clearance by fish from food and water. *Environ. Sci. Technol.* **4**:1203-1213.

Clements, R.G., J.V. Nabhols, D.W. Johnson and **M. Zeeman.** 1993. The use and application of QSARs in the Office of Toxic Substances for ecological hazard assessment of new chemical. In W.G. Landis, J.S. Hughes and M.A. Lewis (eds.), *Environmental Toxicology and Risk Assessment.* ASTM, Philadelphia, PA; Atlantic City, NJ, pp. 56-64.

Cohen, Y., W. Tsai, S.L. Chetty and **G. Mayer.** 1990. Dynamic partitioning of organic chemicals in regional environments: A multi-media screening-level modeling approach. *Environ. Sci. Technol.* **24**:1549-1558.

Cowan, C.E., D.J. Versteeg, R.J. Larson and **P.J. Kloepper-Sams.** 1995. Integrated approach for environmental assessment of new and existing substances. *Reg. Toxicol. Pharm.* **21**:3-31.

Damköhler, G.Z. 1937. Einfluss von Diffusion, Strömung und Wärmetransport. In Eucken-Jakob (ed.), *Der Chemie Ingenieur.* J.W. Edwards, Ann Arbor, MI.

Devillers, J., S. Bintein and **W. Karcher.** 1995. CHEMFRANCE: A regional Level III fugacity model applied to France. *Chemosphere* **3**:457-476.

Devillers, J., P. Boule, P. Vasseur, P. Prevot, R. Steiman, F. Seigle-Murandi, J.L. Benoit-Guyod, M. Nendza, C. Grioni, D. Dive and **P. Chambon.** 1990. Environmental and health risks of hydroquinone. *Ecotoxicol. Environ. Safety* **19**:327-354.

Devillers, J., J. Thioulouse, D. Domine, M. Chastrette and **W. Karcher.** 199 Multivariate analysis of the input and output data in the fugacity model Level I. J. Devillers and W. Karcher (eds.), *Applied Multivariate Analysis in SAR ar Environmental Studies.* Kluwer Academic Publishers, Dordrecht, The Netherlands, p 281-345.

Di Guardo, A., D. Calamari, G. Zanin, A. Consalter and **D. Mackay.** 1994. A fugaci model of pesticide runoff to surface water: Development and validation. *Chemosphe* **28**:511-531.

Droppo, J.G., D.L. Strenge, J.W. Buck, B.L. Hoopes, R.D. Brockhaus, M.B. Walter an **G. Whelan.** 1989. *Multi-media Environmental Pollutant Assessment System (MEPA Application Guidance*, Vol. 1 and 2. Pacific Northwest Laboratories, Richland, WA.

ECETOC. 1994a. *HAZCHEM A Mathematical Model for Use in Risk Assessment Substances.* Special Report N. 8, October 1994. European Centre for Ecotoxicology an Toxicology of Chemicals, Brussels.

ECETOC. 1994b. *Assessment of Non-Occupational Exposure to Chemicals.* Technical Repo No. 58, April 1994. European Centre for Ecotoxicology and Toxicology of Chemica Brussels.

EEC. 1993. Technical Guidance Document in Support of the Risk Assessment Directiv (93/67/EEC) for New Substances Notified in Accordance with the Requirements Council Directive 67/548/EEC. Brussels.

EEC. 1994. Technical Guidance Document in Support of the Commission Regulation (EC N. 1488/94 on Risk Assessment for Existing Substances in Accordance with Counc Regulation (EEC) N. 793/93. Brussels.

FAO. 1990. *Yearbook, Production 1989*, Vol. 43. Food and Agricultural Organization (of th United Nations), FAO Statistics Series No. 94, Rome.

Furtmann, K. 1993. *Phthalate in der aquatischen Umwelt.* LWA-Materialien Nr. 6/9 Landesamt für Wasser und Abfall Nordrhein-Westfalen, Düsseldorf.

Ghetti, P.F., M.P. Marchi and **G.G. Palla.** 1985. Environmental Impact an Ecotoxicological Properties of 3,4-Dichloronitrobenzene (n 63), 2, Dichloronitrobenzene (n 63), 3,4,-Dichloroaniline (n 52), 2,5-Dichloroaniline (n 52 Linuron (n 88), 4-Chloro-2-nitro-aniline (n 27), Propanil (n 104), 2-Amino- chlorophenol. Contract N. 84-3-6600-11-039-11-N. Commission of the Europea Communities, Brussels.

Gobas, F.A.P.C. 1993. A model for predicting the bioaccumulation of hydrophob organic chemicals in aquatic food webs: Application to Lake Ontario. *Ecolo Modelling* **69**:1-17.

Howard, P.H. 1989. *Handbook of Environmental Fate and Exposure Data for Organic Chemica* (4 Vols.). Lewis Publishers, Boca Raton, FL.

Hunter, B. 1992. *The Statesman's Year-book. Statistical and historical annual of the states of th world for the year 1992-1993.* Macmillan Press, London.

Iman, R.L. and **J.C. Helton.** 1988. An investigation of uncertainty and sensitivity analysis techniques for computer models. *Risk Analysis* **8**:71-90.

IARC. 1982. *Some industrial chemicals and dyestuffs.* Monograph Vol. 29. International Agency for Research on Cancer, Lyon.

IPCS. 1992. *Environmental Health Criteria 131, Diethylhexyl phthalate.* International Programme on Chemical Safety, Geneva.

Jongbloed, R.H., J. Pijnenburg, B.J.W.G. Mensink, T.P. Traas and **R. Luttik.** 1994. A Model for Environmental Risk Assessment and Standard Setting Based on Biomagnification. Top Predators in Terrestrial Ecosystem. RIVM Report n. 719101012, Bilthoven, The Netherlands.

Jørgensen, S.E. 1990. *Modelling in Ecotoxicology.* Elsevier, Amsterdam.

Jørgensen, S.E. 1994. *Fundamentals of Ecological Modelling.* Elsevier, Amsterdam.

Jori, A., D. Calamari, A. Di Domenico, C.L. Galli, E. Galli, M. Marinovich and **V. Silano.** 1986. Ecotoxicological profile of xylenes. *Ecotoxicol. Environ. Safety* **11**:44-80.

Jury, W.A., D. Russo, G. Streile and **H. El Abd.** 1990. Evaluation of volatilization by organic chemicals residing below the soil surface. *Water Resources Res.* **26**:13-20.

Jury, W.A., W.F. Spencer and **W.J. Farmer.** 1983. Behavior assessment model for trace organic in soil: I model description. *J. Environ. Qual.* **12**:558-566.

Lyman, W.J., W.F. Reehl and **D.H. Rosenblatt.** 1982. *Handbook of Chemical Property Estimation Methods.* McGraw-Hill, New York.

Mackay, D. 1989. An approach to modeling the long term behavior of an organic contaminant in a large lake: Application to PCBs in Lake Ontario. *J. Great Lakes Res.* **15**:283-297.

Mackay, D. 1991. *Multimedia Environmental Fate Models: The Fugacity Approach.* Lewis Publishers, Chelsea, MI.

Mackay, D., S. Paterson and **W.Y. Shiu.** 1992. Generic models for evaluating the regional fate of chemicals. *Chemosphere* **24**:695-717.

Mackay, D., S. Paterson, G. Kicsi, A. Di Guardo and **C.E. Cowan.** 1995a. Assessing the fate of new and existing chemicals: A five stage process. *Environ. Toxicol. Chem.* (in review).

Mackay, D., S. Paterson, A. Di Guardo and **C.E. Cowan.** 1995b. Evaluating the environmental fate of a variety of types of chemicals using the EQC model. *Environ. Toxicol. Chem.* (in review).

Mackay, D., S. Paterson, D.D. Tam, A. Di Guardo and **D. Kane.** 1995c. ChemCAN: A regional Level III fugacity model for assessing chemical fate in Canada. *Environ. Toxicol. Chem.* (in review).

Mackay, D., S. Paterson, G. Kicsi, C.E. Cowan, A. Di Guardo and **D. Kane.** 1995d. Assessment of chemical fate in the environment using evaluative, regional and local-scale models: Illustrative application to chlorobenzene and linear alkylbenzene sulfonates. *Environ. Toxicol. Chem.* (in review).

Mackay, D., W.Y. Shiu and **K.C. Ma.** 1992a. *Illustrated Handbook of Physical-Chemical Properties and Environmental Fate for Organic Chemicals,* Vol I. CRC Press, Boca Raton, FL.

Mackay, D., W.Y. Shiu and **K.C. Ma.** 1992b. *Illustrated Handbook of Physical-Chemical Properties and Environmental Fate for Organic Chemicals,* Vol II. CRC Press, Boca Raton, FL.

Mackay, D., W.Y. Shiu and **K.C. Ma.** 1994. *Illustrated Handbook of Physical-Chemical Properties and Environmental Fate for Organic Chemicals,* Vol III. CRC Press, Boca Raton, FL.

Mackay, D., W.Y. Shiu and **K.C. Ma.** 1995. *Illustrated Handbook of Physical-Chemical Properties and Environmental Fate for Organic Chemicals,* Vol IV. CRC Press, Boca Raton, FL.

McCuen, R.H. 1973. The role of sensitivity analysis in hydrologic modeling. *J. Hydrol.* **18**:37-53.

McKone, T.E. 1993. CalTOX, A Multi-media Total-Exposure Model for Hazardous Wastes Sites Part II: The Dynamic Multi-media Transport and Transformation Model. A report prepared for the State of California, Department Toxic Substances Control by the Lawrence Livermore National Laboratory No. UCRL-CR-111456PtII. Livermore, CA.

OECD. 1991. *Background Document 2: Local and Global Models. For the OECD workshop on the application of simple models for environmental exposure assessment.* OECD, Paris.

OECD. 1994. *US EPA and EC Joint Project on the Evaluation of (Quantitative) Structure-activity Relationships.* No. 88. OECD Environmental Monograph. OECD, Paris.

Oreskes, N., K. Shrader-Frechette and **K. Belitz.** 1994. Verification, validation, and confirmation of numerical models in the earth sciences. *Science* **263**:641-646.

Rapaport, R.A. and **W.S. Eckhoff.** 1990. Monitoring LAS in the environment: 1973-1986. *Environ. Toxicol. Chem.* **9**:1245-1257.

Reible, D.D., K.T. Valsaraj and **L.T. Thibodeaux.** 1991. Chemodynamic Models for Transport of Contaminants From Sediment Beds. In O. Hutzinger (ed.), *Handbook of Environmental Chemistry,* Vol. 2, Part F. Springer Verlag, Heidelberg, p. 185-228.

Scheil, S., G. Baumgarten, B. Reiter, S. Schwartz, J.-O. Wagner, M. Matthies and **S. Trapp.** 1994. CemoS - Eine objektorientierte Software zur Expositionsmodellierung. In K. Totsche, M. Matthies *et al.*(eds.), *Ecoinforma 1994*, Vol. 7 391-404.

Scheil, S., G. Baumgarten, B. Reiter, S. Schwartz, J.-O. Wagner, S. Trapp and **M. Matthies.** 1995. An object-oriented software for fate and exposure assessments. *Environ. Sci. Pollut. Res.* (In review).

Simonich, F.L. and **R.A. Hites.** 1994. Importance of vegetation as a route by which Polycyclic Aromatic Hydrocarbons are removed from the atmosphere. *Nature* **370**:49-51.

Suedel, B.C., J.A. Boraczek, R.K. Peddicord, P.A. Clifford and **T.M. Dillon.** 1994. Tropic transfer and biomagnification potential of contaminants in aquatic ecosystems. *Rev. Environ. Cont. Toxicol.* **136**:21-89.

Swackhamer, D.L. and **R.S. Skoglund.** 1993. Bioaccumulation of PCBs by algae: Kinetics versus equilibrium. *Environ. Toxicol. Chem.* **12**:831-838.

Thibodeaux, L.J. 1979. *Chemodynamics*. Wiley Interscience, New York.

Trapp, S., C. McFarlane and **M. Matthies.** 1994a. Model for uptake of xenobiotics into plants: Validation with bromacil experiments. *Environ. Toxicol. Chem.* **13**:413-422.

Trapp, S., M. Matthies and **A. Kaune.** 1994b. Transfer von PCDD/F und anderen organischen Umweltchemikalien im System Boden-Pflanze-Luft. I. Modellierung des Transferverhaltens. UWSF - Z. Umweltchem. *Ökotox.* **6**:31-40.

Trapp, S. and **C. McFarlane** (eds.). 1995. *Plant Contamination Modeling and Simulation of Organic Chemical Processes*. CRC Press, Boca Raton, FL.

UK-MAFF. 1987. Survey of plasticiser levels in food contact materials and in foods. Food Surveillance Paper No. 21. Ministry of Agriculture, Fisheries and Food, London.

USEPA. 1992. Guidelines for Exposure Assessment, *Federal Register* **57**:22889-22938.

van de Meent, D. 1993. SimpleBOX: A Generic Multi-Media Fate Evaluation Model. RIVM Report No. 6727200001. Bilthoven, The Netherlands.

van de Meent, D. and **J.H.M. de Bruijn.** 1995. A modeling procedure to evaluate the coherence of independent derived environmental quality objectives for air, water and soil. *Environ. Toxicol. Chem.* **14**:177-186.

van der Zandt, P.T.J. and **C.J. van Leeuwen.** 1992. A proposal for priority setting of existing chemical substances. Ministry of Housing, Physical Planning and the Environment, The Netherlands.

Wagner, J.-O. and **M. Matthies.** 1995. Guidelines for selection and application of fate and exposure models: Model criteria and model data bank. *Environ. Sci. Pollut. Res.* (submitted).

Wania, F. and **D. Mackay.** 1995. A global distribution model for persistent organic chemicals. *Sci. Total Environ.* **160/161**:211-232.

Young, D.R., M.D. Moore, T.-K. Jan and **R.P. Eganhouse.** 1981. Metals in seafood organisms near a large California municipal outfall. *Marine Pollut. Bull.* **12**:134-138.

Young, D.R. 1988. Report on the Assessment and Application of Pollutant Biomagnification Potential in Near Coastal Waters. ERL-N Contribution No. N-065, Pacific Division Environmental Research Laboratory, Office of Research and Development. USEPA, Newport, OR.

SETAC

A Professional Society for Environmental Scientists and Engineers and Related Disciplines Concerned with Environmental Quality

The Society of Environmental Toxicology and Chemistry (SETAC), with offices in North America and Europe, is a nonprofit, professional society that provides a forum for individuals and institutions engaged in the study of environmental problems, management and regulation of natural resources, education, research and development, and manufacturing and distribution.

Goals

- Promote research, education, and training in the environmental sciences
- Promote systematic application of all relevant scientific disciplines to the evaluation of chemical hazards
- Participate in scientific interpretation of issues concerned with hazard assessment and risk analysis
- Support development of ecologically acceptable practices and principles
- Provide a forum for communication among professionals in government, business, academia, and other segments of society involved in the use, protection, and management of our environment

Activities

- Annual meetings with study and workshop sessions, platform and poster papers, and achievement and merit awards
- Monthly scientific journal, *Environmental Toxicology and Chemistry*, SETAC newsletter, and special technical publications
- Funds for education and training through the SETAC Scholarship/Fellowship Program
- Chapter forums for the presentation of scientific data and for the interchange and study of information about local concerns
- Advice and counsel to technical and nontechnical persons through a number of standing and *ad hoc* committees

Membership

SETAC's growing membership includes more than 3,600 individuals from government, academia, business, and public-interest groups with technical backgrounds in chemistry, toxicology, biology, ecology, atmospheric sciences, health sciences, earth sciences, and engineering.

If you have training in these or related disciplines and are engaged in the study, use, or management of environmental resources, SETAC can fulfill your professional affiliation needs. Membership categories include Associate, Student, Senior Active, and Emeritus.

For more information, contact SETAC, 1010 North 12th Avenue, Pensacola, Florida, 904-469-1500.